Handwritten note: When using 100x / field iris diaphragm to its / smallest diameter

Delafield H&	Heidenhain (Hematoxylin)	Tissue			Safranin–Fast Green (S–FG)
Paraffin sections (mouse tissues and *Tilia*)					
Frozen sections (mouse tissues)					
Celloidin sections (*Tilia*); paraffin due					
Frozen sections due	Stain (mouse tissues and onion root tip)	Set up onions for root tips			
		Hydra and onion root tip			
Celloidin sections due	Slides due		Stain (mouse tissues and onion root tip)		
		Slides due		Stain (mouse tissues & *Tilia*)	
			Slides due		Stain (mouse tissues & *Tilia*)

Microtechniques: A Laboratory Guide

Ruth L. Willey
University of Illinois at Chicago Circle

The Macmillan Company, New York

Collier-Macmillan Limited, London

The Macmillan Company
866 Third Avenue, New York, New York 10022

Collier-Macmillan Canada, Ltd., Toronto, Ontario

Library of Congress catalog card number: 75-114330

PRINTING   23456789   YEAR   3456789

PREFACE

This manual is designed to be used in a standard college course in microtechniques; it is not intended to be an encyclopedia of histological methods. I have tried to include, along with diversified subject matter, different types of preparative techniques and stain methods. By the end of the course all participants should be able to handle independently any of the many methods of tissue preparation for the plant and animal sciences by using the light microscope. The schedule is designed for a ten-week quarter. However, several alternative methods, especially the whole mounts, have been added to accommodate a fifteen-week semester course and to provide teachers with choices and ideas for special projects.

The general philosophy of the course is based on the conviction that reliable diagnostic histology and cytology are impossible when methodology is routine. Even a laboratory technician who follows the same routine every day must have sufficient knowledge of the processes involved to recognize and correct the problems that constantly occur. In addition, each student, when faced with a problem in the biological sciences, must have sufficient background in the available types of microtechniques to be able to choose the best method for his particular experiment. Students should be able to judge accurately the quality and reliability of a preparation as well as to produce neatly labeled, well-stained slides. This necessitates critical examination by each student of his own slides, and the last two weeks of slide analysis often prove to be the most valuable part of the course. One could call it the "moment of truth." Included in Chapter 7 is a list of questions which each student must answer from the evidence of his own slides. After all, there is a good reason for the variety of methods; each method should have a specific purpose and each student must learn to choose intelligently on the basis of that purpose. Also, because few students in the early years of their biological education have an adequate knowledge of cell anatomy, sections on cell ultrastructure and cell anatomy as a correlative background for the histological analysis are included.

All the techniques described in this manual have been used in the laboratory by students and generally work under even the most difficult conditions. However, experience has shown that no technique can be guaranteed because, for many unknown and seemingly inconsequential reasons, methods that work for one student will not necessarily work the same way for another student. One of the major points to be stressed is the necessity of continual and intelligent modification of any set of directions. This requires that each student understand exactly what he is doing rather than blindly follow a cookbook recipe. In the past many students have learned the hard way that a technique does not necessarily work to perfection on the first attempt.

Few collections of techniques arise as a single effort of one person, and I am indebted to many colleagues for ideas and suggestions. I am especially indebted to Dr. R. B. Willey of the University of Illinois, Chicago, for his constant advice and help, to Dr. Alexander Wetmore of Harvard University for advice concerning the paraffin and celloidin schedules for plant material, and to Mr. Paul Conant and Dr. Carl Hagquist of Triarch Products, Inc. Dr. R. B. Willey and Mrs. R. Dyer were instrumental in the development of the maceration technique for *Hydra*. Miss S. Stamler worked out the final protocol for the onion root tip squash. The following students were responsible for the preparations photographed for the illustrative plates: R. Evenhouse, G. Hoffman, D. Kerrigan, M. Robin, G. Simone, and F. Slansky. The technical assistance of Mr. David Mucha and Mrs. Esther Kerster was essential to the final completion of this manual.

RLW

CONTENTS

Microtechniques: A Laboratory Guide

Chapter 1

TISSUE-
PROCESSING DIRECTIONS

ORGANS TO BE FIXED

Mouse Tissues

Liver
Testis
Heart
Duodenum (cross section)

Plant Tissues

Tilia twig (cross, radial, tangential sections)
Onion root tip (longitudinal section)

CHART OF TISSUES AND FIXATIVES
(The recommended number of pieces of tissue per *pair* of students)

Tissue	Carnoy	Bouin	Helly	FAA
Liver (Lv)	4	2	2	0
Testis (Ts)	2	1	1	0
Heart (Ht)	2	1	1	0
Duodenum (Du)	4	2	2	0
Tilia twig (Ti)	0	0	0	4
Onion root tip (O)	0	0	0	4

EMBEDDING MEDIUM AND SECTION THICKNESS

Tissuemat

All mouse tissues, onion root tips, *Tilia* twig (two pieces)—10 microns

Parlodion

Tilia twig (two pieces)—15 microns

Orientation of Tissues for Sectioning

Duodenum (cross section)
Tilia (cross, radial, and tangential sections; one each on a slide, see Fig. 2–3)
Onion root tip (longitudinal section)

METHYLENE BLUE–PHLOXINE

	Bouin				Helly				Carnoy			
Stain	Lv	Ts	Du	Ht	Lv	Ts	Du	Ht	Lv	Ts	Du	Ht
Methylene blue	x				x				x			
Phloxine	x				x				x			
Methylene blue and phloxine together	x	x	x	x	x	x	x	x	x	x	x	x
4°C pretreated									x	x	x	x
60°C pretreated									x	x	x	x

DELAFIELD'S HEMATOXYLIN AND EOSIN Y (PARAFFIN)

	Bouin				Helly				Carnoy				FAA
Stain	Lv	Ts	Du	Ht	Lv	Ts	Du	Ht	Lv	Ts	Du	Ht	Ti
Hematoxylin and eosin Y	x	x	x	x	x	x	x	x	x	x	x	x	x

DELAFIELD'S HEMATOXYLIN AND EOSIN Y (FROZEN)

Mouse liver (1 section per slide)

DELAFIELD'S HEMATOXYLIN AND EOSIN Y (CELLOIDIN)

Tilia twig (cross, radial, and tangential sections on a single slide)

HEIDENHAIN'S IRON HEMATOXYLIN

Stain	Bouin				Helly				Carnoy				FAA
	Lv	Ts	Du	Ht	Lv	Ts	Du	Ht	Lv	Ts	Du	Ht	O
Hematoxylin and orange G	x	x	x	x	x	x	x	x	x	x	x	x	x

FEULGEN AND FAST GREEN FCF

Stain	Carnoy				FAA
	Lv	Ts	Du	Ht	O
Schiff reagent (without hydrolysis) and fast green		x			x
Feulgen and fast green	x	x	x	x	x

PERIODIC ACID–SCHIFF

Stain	Carnoy				FAA
	Lv	Ts	Du	Ht	Ti
Schiff reagent (without hydrolysis) and hematoxylin	x	x	x	x	
Periodic acid–Schiff and hematoxylin	x	x	x	x	x
Periodic acid–Schiff and hematoxylin after pretreating with diastase	x	x	x	x	

SAFRANIN AND FAST GREEN FCF

	FAA	
Stain	Ti	O
Safranin and fast green FCF	x	x

Chapter 2

GENERAL ROUTINE
FOR
PREPARATION OF TISSUES

A. FIXATION

The death of a cell leads rapidly to a breakdown of cell constituents resulting from internal enzymes and from damage caused by external agents such as bacteria and enzymes. The cell contents begin to diffuse out of position as well as to change chemically. The aim of tissue preparation for microscopic investigation is the preservation of the cell in a state as close to the living condition as possible. Therefore, the selected tissue must be placed in the selected fixative as soon as possible after death—there must be no delay. All solutions must be ready before the tissue is removed from the living state. The fixative solution should be selected for

1. Rapid penetration
2. Ability to change the soluble cell contents to insoluble substances that will remain in their original position
3. Protect tissues against shrinkage and distortion during subsequent treatment
4. Constituents enabling the cell to be seen through the light microscope by changing the refractive index of the cell organelles or by making them stainable

Primary fixatives, solutions of a single chemical, seldom have all the properties of a good fixative. Usually mixtures are used that take advantage of the specific action of each single compound to make a fixative mixture that will fulfill all the requirements (see Appendix A).

B. FIXATIVES

Ideally the choice of a fixative will depend on the properties of the tissue plus the stain or reaction to be used. However, some fixatives are good enough for routine work, and special fixatives need be applied only for specialized reactions of tissues.

1.   **BOUIN'S FLUID**

> Picric acid, saturated aqueous solution (1.2 gm/100 cc)   75 cc
> Formalin (40 percent formaldehyde)   25 cc
> Glacial acetic acid   5 cc

a.   Fix 12 to 24 hours depending on tissue size.
b.   Place directly in 70 percent ethanol with a small amount of lithium carbonate for at least 24 hours. Change the ethanol several times during this period.

As much of the picric acid as possible should be removed at this time because it tends to slow down paraffin penetration during embedding. Small amounts of lithium carbonate added to the ethanol help to remove the picric acid. All picric acid must be removed before staining is begun.

Bouin's fluid works well with most mammalian and invertebrate tissues. It is good for nuclei, but fixes cytoplasmic structures only indifferently. It is an excellent field fixative. The solution keeps indefinitely and tissues can be left in it for weeks.

2.   **ZENKER'S OR HELLY'S FLUID**

Stock

> Potassium dichromate   2.5 gm
> Mercuric chloride   5.0 gm
> Sodium sulfate   1.0 gm
> Distilled water   100.0 cc

Zenker's Fluid

Immediately before use, add 1 part (5 cc) glacial acetic acid to 19 parts (95 cc) stock solution; use *immediately*.

Helly's Fluid

Immediately before use, add 1 part formalin (40 percent formaldehyde) to 10 parts stock solution; use *immediately*.

a.   Fix 12 to 24 hours (depending on the size of the tissue) *in the dark*.
b.   Wash in running tap water for 24 hours.
c.   Transfer to 50 percent ethanol for 2 hours; then transfer to 70 percent ethanol.

d. To remove mercuric chloride, place the tissues in 70 percent ethanol to which enough iodine solution has been added to give a deep brown color. Leave in iodine-70 percent ethanol until the color fails to disappear (up to 6 hours).

Iodine Solution

Potassium iodide	3 gm
95 percent ethanol	100 cc
Dissolve and add	
Iodine	2 gm

These are excellent fixatives. Helly's is especially good for cytoplasmic granules (secretion, neurosecretion). Helly's does not harden tissues as much as Zenker's, but Zenker's penetrates faster. Stock solution keeps indefinitely; keep it away from strong sunlight.

3. CARNOY'S FLUID

Glacial acetic acid	10 cc
Absolute ethanol	60 cc
Chloroform	30 cc

a. Fix 20 minutes to 3 hours depending on the size of the tissue. Tissue should be kept small for this fixative—no more than 5 mm across.
b. Transfer to 95 percent ethanol (1 hour) and then to 70 percent ethanol for temporary storage or dehydrate immediately and embed in paraffin (see Section D)—preferably the latter.

Carnoy's works very rapidly and is an excellent cytoplasmic fixative. It is used extensively for histochemical work. The fixative keeps indefinitely.

Mahdissan's Fluid is a variant of Carnoy's Fluid. Some laboratories use it extensively for insect tissues.

Absolute ethanol	60 cc
Chloroform	30 cc
Formalin (40 percent formaldehyde)	10 cc
Glacial acetic acid	5 cc

4.   BAKER'S FORMAL-CALCIUM FLUID

Formalin (40 percent formaldehyde)	10 cc
Calcium chloride (anhydrous)	1 gm
Distilled water	90 cc
Calcium carbonate	excess

(Keep excess carbonate in the bottom of the bottle.)

a. Fix 2 to 3 days. Store in fluid if necessary.
b. If lipid preservation is not important, transfer the tissue directly to 70 percent ethanol and treat it the same way as other tissues to be embedded in paraffin. Tissues may be stored in 70 percent ethanol.

When lipids are fixed special methods are required for sectioning and staining. (See frozen section references.) Avoid fat solvents (xylene and benzene). Because Baker's fluid is neutral, it is also used for preservation of calcareous structures. However, it should not be used for localization of phosphatases. Penetration is very slow; therefore it is important to cut the tissues into small pieces. Formalin fluids are excellent for preservation of an entire organism if it is to be used for general dissection. The fluid keeps indefinitely.

5.   REGAUD'S SOLUTION (Conn *et al.*, 1960)

Potassium dichromate	2.4 gm
Distilled water	80.0 cc
Formalin	20.0 cc

a. Fix, 2 days, change the fluid every day.
b. 3 percent potassium dichromate, 4 to 8 days, change the fluid every day.
c. Running water, 24 hours.
d. Transfer to 50 percent ethanol for 2 hours, then transfer to 70 percent ethanol.

Regaud's solution is used for mitochondria. Mix the solution immediately before use. Fix the tissues in the dark.

6.   FAA SOLUTION

95 percent ethanol	50 cc
Glacial acetic acid	5 cc
Formalin	10 cc
Distilled water	35 cc

a. Fix thin leaf pieces 12 hours, small twigs 24 hours.
b. Wash in 2 to 3 changes of 50 percent ethanol, 24 hours each.
c. Store temporarily in 70 percent ethanol. Long-term storage (1 month or more) should be in 1 percent glycerine in 70 percent ethanol.

This is an excellent, fast-penetrating fixative for plants. Tissues may be stored indefinitely in FAA. It is an excellent field fixative.

7. **CRAF SOLUTION** (Courtesy of Triarch Products, Inc.)

Stock

Chromium trioxide	1.0 gm
Distilled water	96.0 cc
Glacial acetic acid	4.0 cc

Immediately before use, mix equal parts (1:1) of stock solution and 25 percent formalin.

a. Fix small, soft tissues (onion root tip) for 24 hours, larger tissues for 3 days to 1 week.
b. Wash in running water 6 hours to overnight.
c. Dehydrate in the following successive solutions: 15, 30, and 50 percent ethanol; 1/2 hour (root tip) to 4 hours (larger tissue) each.
d. Store in 70 percent ethanol until ready to process.

For subsequent treatment of animal tissues in paraffin, see Section D (General Paraffin Routine) and for plant tissues, see Section E (Specific Paraffin Routine for Plant Tissues). Directions to embed plant or animal tissues in celloidin are in Section F (Celloidin Embedding Routine).

C. DISSECTION AND FIXATION OF MOUSE TISSUES

Tissues must be removed from the mouse and placed in the fixative as rapidly as possible. Speed is essential. Have your dissection equipment ready and at hand. Plan ahead; have all the solutions made up before you kill the mouse. Have enough fixative so that each piece of tissue will be covered by at least ten times its own volume (about 100 cc of each fixative if you are following the Tissue-Processing Directions). Fully label all jars (and capsules, if used) with pencil or indelible ink ahead of time.

The mouse must be killed as quickly and humanely as possible. Place it in a large jar with a tight lid into which you have previously placed paper towels soaked with ether or chloroform. Place a dry towel on top of the others to prevent the mouse's coming into contact with the liquid. As soon as the mouse stops breathing, remove it and cut the spinal cord just behind the head. If you are unsure about the internal anatomy of a mouse, you should work with a partner who can dissect a mouse quickly. Place the body on its back on a paper towel, and moisten the fur along the ventral midline with water. Pinch up the skin of the lower abdomen with forceps. Cut through the skin and then the abdominal muscles with scissors. With the tip of the scissors make a longitudinal cut anteriorly along the midventral line up to and through the rib cage. Keep the tip of your scissors near the abdominal muscles to prevent accidental cutting into the abdominal organs.

Germinal cells and cells with high enzyme content degenerate notoriously fast at death. Therefore, it is necessary to dissect out the tissues in the following order:

1. Liver. Cut pieces no more than 5 mm in diameter from the edge of the liver. Rinse them quickly in saline solution (0.7 percent sodium chloride) and place each in its proper fixative.
2. Duodenum. Push the liver aside, pull up the stomach, and cut the duodenum into lengths no more than 3 mm long. Rinse them quickly in the saline solution. Flush out the lumen gently with a pipette filled with saline solution and place each piece in its proper fixative.
3. Testis. Slit open the scrotal sacs with a pair of scissors. Remove the entire testis and drop it directly into the fixative. Now is not the time to find out that you killed a female mouse.
4. Heart. Remove the heart, cut it in half longitudinally, and rinse it vigorously in saline solution to remove blood clots. Place each piece in its proper fixative.

When trimming the tissues, handle them with care. Do not pinch or stick them with forceps. Trim them with a razor blade or sharp scissors. Allow only stainless steel tools to come into contact with mercuric chloride fixatives, which are very corrosive. No tissue should have a diameter greater than 5 mm. When the testis has been slightly hardened by the fixative (about 5 minutes), slice it open carefully with a *sharp* razor blade. The testis is a very soft tissue and crushes easily.

Labeling of each tissue is important because of the necessity of identification during the many steps in subsequent preparation. Label each piece with the kind of tissue and the fixative. Keep tissues in separate, labeled vials, or wrap them in cheesecloth and attach a label with string, or place them in a small perforated capsule. Tissues that must be washed in running water will have to be placed in cheesecloth packets or in commercially available capsules. When many tissues are being processed, large numbers of suitable containers are needed. Baby food jars are available in inexhaustible supply, have acid-resistant screw caps, and are disposable (especially the jars that contained mercuric chloride solutions). Instruments and glassware contaminated with mercuric chloride must never be allowed to come into contact with living tissue.

D. GENERAL PARAFFIN ROUTINE

FIXATION

The choice of a fixative and the length of time necessary for fixation will depend on the tissue and the desires of the technician.

DEHYDRATION

A standard graded series of alcohols are used to dehydrate or hydrate tissues to prevent shrinkage or distortion of the cells when one kind of solvent must be replaced by another solvent.

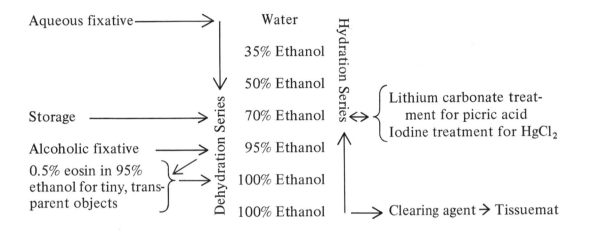

One to two hours in each solution should be adequate. To ensure complete removal of water during dehydration, use two changes of 100 percent ethanol of 1 hour each. Never leave tissues in 95 or 100 percent ethanol more than a total of 2 hours or the tissues will harden. Tissues should be stored in 70 percent ethanol during any interruption in routine.

CLEARING

1. 100 percent ethanol : cedarwood oil (1:1), 1 to 2 hours.
2. Fresh cedarwood oil; overnight or longer to clear. Tissues can be left in cedarwood oil indefinitely. It does not harden the tissue.

There are many clearing agents in use. Often the choice depends on the whim of the technician. It is desirable to use an agent that does not harden the tissue, that clears properly, and that is miscible with embedding paraffin as well as with ethanol. The clearing agent serves as a transition medium between two immiscible compounds, ethanol and paraffin.

a. Benzene or toluene. Commonly used; clears overnight.
b. Cedarwood oil. Slightly slower in penetrating than benzene is; does not cause hardening; does not interfere too seriously with paraffin penetration if it is not completely removed.
c. Methyl benzoate. Very good for clearing; does not harden tissues; avoids use of xylene; benzene is used to rinse out the clearing agent. It penetrates almost as fast as does cedarwood oil (12 to 24 hours) and is most valuable with the Peterfi celloidin-impregnation technique.
d. Dioxan. Many directions and techniques make use of dioxan, which is miscible both with water and paraffin. It is used primarily when time is important because the tissues may be embedded with paraffin within 4 hours after fixation. The tissues are transferred to dioxan straight from Bouin's fluid or a formalin fixative. The dioxan is changed 3 times within 4 hours and the tissues are transferred directly to paraffin (3 changes are made in a total of 90 minutes). Dioxan causes greater shrinkage than xylene does. In addition, it is *dangerous.* Fumes of dioxan are toxic to humans; reportedly it is a liver poison. Dioxan *must* be used in a hood at all times and must be stored in tightly sealed jars. Frankly, it is not worth it.

EMBEDDING

1. Rinse in xylene for 2 or 3 minutes to remove any excess clearing agent. If benzene was used as a clearing agent, use benzene instead of xylene.
2. Transfer to a mixture of xylene (benzene) saturated with Tissuemat chips which rests on top of the paraffin oven or in some warm spot at about 35°C, 15 minutes.
3. Infiltrate with hot Tissuemat in the 58°C paraffin oven in an open container in 4 changes with the following time sequence: 30 minutes, 30 minutes, 1 hour, 30 minutes.

Never leave a tissue in the paraffin oven for more than 4 hours. The shorter the time in the hot oven with adequate paraffin impregnation and evaporation of the clearing agent, the better for the tissue. Tissues become increasingly harder and more brittle as they are heated. Generally, 56 to 58°C Tissuemat is used for embedding. During the winter 54 to 56°C Tissuemat may be used if the tissue is cut in a cool room. During the summer it may be necessary to use 60 to 63°C Tissuemat. This is to be avoided if possible in order not to "cook" the tissue. "Cooked" tissue does not stain well and most details are destroyed. Bioloid paraffin (a mixture of paraffin and other waxes) and Paraplast (paraffin and plastic polymers) are also common embedding media.

4. Pour melted Tissuemat (fresh) into a paper box (Fig. 2–1) or plastic ice-cube cup. Cool the bottom of the container in a bowl of cold (ice) water, but keep the top surface liquid. Heat forceps in an alcohol flame and draw the tips through any top scum to keep it melted. Transfer the tissue with hot forceps and orient it properly in one end or a corner of the box. (Use a paper label to mark the point of orientation.) Cool the top surface of the Tissuemat by blowing gently on it. Tissues at this stage are very brittle; handle them with care. As soon as a scum of Tissuemat has formed on top, sink the cup gently into the cold water. The block will be ruined if it is submerged before its upper surface has formed a protective scum. Cool thoroughly in cold running tap water. If you use ice water for the final cooling, you are apt to split the block owing to too rapid shrinkage. Tissuemat naturally splits in the line of least resistance—right through the tissue. If you use plastic ice-cube cups, the Tissuemat block can be removed easily as soon as it is cooled and the cup can be used again.

Orientation of tissues in the Tissuemat block is important for tissues such as duodenum when sections in a predetermined plane, such as cross sections, are required. Also, trimming is excessively difficult in a block embedded with two or more tissues if they are not carefully lined up before the Tissuemat is cooled.

CUTTING SECTIONS AND FIXING THEM ON SLIDES

1. Initially trim the Tissuemat block to a size that will fit the face of a 5/8 inch square wooden cube.
2. Label the side of the wooden block and melt a Tissuemat chip on the end surface of the wooden block. Press the Tissuemat block into the molten Tissuemat before it cools. Then the combined block is cooled thoroughly in cold water. Tissues may be stored indefinitely in Tissuemat. Trim the face of the block to a size in which the tissue is surrounded by about a 1 mm frame of paraffin. Trim the upper and lower edges *straight* and *parallel.* Clamp the wooden block in the microtome carrier.
3. Cut sections on the microtome; 10 microns. As they are cut place the ribbons on sheets of clean paper in clean boxes with covers (see Chapter 3, A).
4. Use *clean* glass slides. With a carborundum or diamond-tipped pencil, engrave one end of the slide to identify the tissue and fixative. Slides can be cleaned with Bon Ami, rinsed thoroughly in distilled water, dipped several times in 95 percent ethanol, and either drained dry or wiped with a piece of clean, lintless cloth.
5. Turn the slide over and spread a small drop of Mayer's albumen adhesive (Appendix B) evenly over the surface. Spread the adhesive thinly and wipe off the excess with a clean finger. The scratches of the engraved identification which are now on the bottom of the slide can be located with a fingernail when the question arises about which side of the slide has the section.

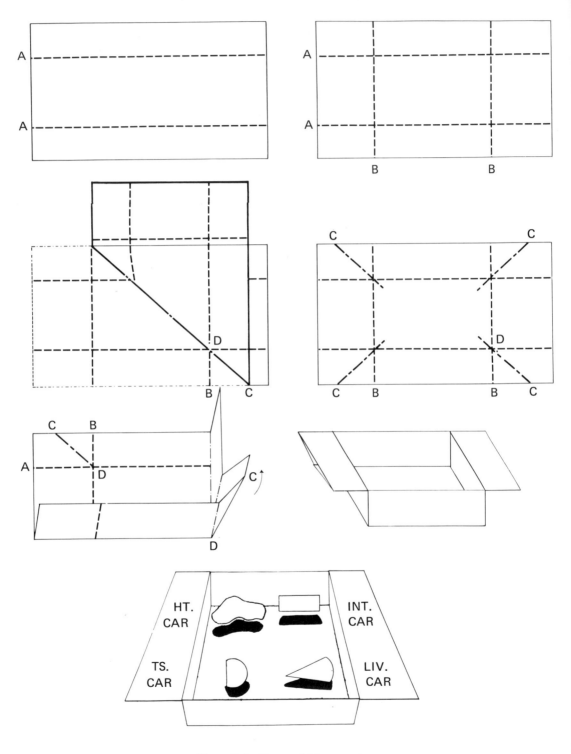

Figure 2-1.   How to fold a paper box.

6. Flood slide with distilled water. Place two or three sections on the water surface near the center of the slide. Paraffin expands about 20 percent so do not take too many sections. Remember that they must be covered by a 22 X 22 mm coverslip.

7. Heat gently until the sections spread and flatten out. Do not melt the Tissuemat. The slide can be heated over a small alcohol flame or on a warm spreading table (40 to 45°C).

8. Drain off the excess water with a paper towel or tissue; do not drain off any sections with it. Place the sections in the center of the slide in a line perpendicular to the long axis of the slide.

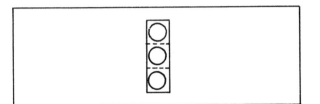

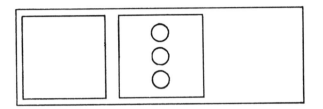

Figure 2-2. Placement of sections and a label on a slide.

9. Dry the slides on a warm table for at least 1 hour before staining. Store them preferably overnight in a slide box in the 45°C oven. After that, store at room temperature. Always keep slides covered to keep them clean.

E. SPECIFIC PARAFFIN ROUTINE FOR PLANT TISSUES

FIXATION

Use fresh material and fix it in FAA as soon as possible after it has been collected. Cut it into small pieces with a sharp razor blade to avoid bruising the tissues and to promote rapid fixation. Onion root tips should be 5 mm long. The rate of fixative penetration can be increased by fixation under a vacuum (–15 lb per square inch). All tissues should have sunk to the bottom of the bottle by the end of fixation.

DEHYDRATION

Solutions 1 to 2: 2 (root tip) to 4 (twig) hours each
Solutions 3 to 5: 4 (root tip) to 6 (twig) hours each

N–BUTYL ALCOHOL SERIES

	1	2	3	4	5
Water	30	15	5	0	0
Butyl alcohol	20	35	55	75	100
Ethyl alcohol	50	50	40	25	0

Make three changes of 100 percent butyl alcohol. For storage of a month or more, run the tissues into 70 percent ethyl alcohol and store in 1 percent glycerin in 70 percent ethyl alcohol.

INFILTRATION

1. Transfer the tissues to a corked vial with shavings of 52°C Tissuemat in 100 percent butyl alcohol. Leave the material 1 day at room temperature.
2. Place the corked vial on top of a 45°C oven for 2 days. Add more Tissuemat on the second day.
3. Remove the cork and place the open vial in a 45°C oven for 2 days. Add more 52°C Tissuemat as it is needed.
4. Place the open vial in a 58°C oven for 1 day. Add more 52°C Tissuemat if the previous chips have dissolved.
5. Pour off 25 percent of the Tissuemat-alcohol mixture. Add melted 56°C Tissuemat and let the material stand for 2 hours.
6. Pour off 50 percent of the mixture and replace it with melted 56°C Tissuemat, leave it for 2 hours.
7. Replace with two complete changes of 56°C Tissuemat, 2 hours for the first, overnight for the second.
8. Fresh 56°C Tissuemat for 3 to 5 days.
9. Embed in 56°C Tissuemat.

SECTIONING

Dense plant material will often cut more easily if the trimmed blocks are soaked in water (add a small crystal of thymol) for 1 to 3 days. For sections less than 10 microns, it may be necessary to chill the face of the block and the knife with ice.

SPREADING

Smear slides with Haupt's gelatin adhesive (Appendix B). Float the sections on 2 percent formalin. Dry the slides overnight before you stain them.

F. CELLOIDIN EMBEDDING ROUTINE

Celloidin is used as an embedding matrix for large objects (cat brain), very hard objects (wood, bone), and delicate objects that might be damaged by heat (small coelenterate medusae and eggs filled with yolk). The celloidin procedure takes longer than the standard paraffin procedure and it is very difficult to cut thin or serial sections. Generally this procedure is used only if paraffin is unsuitable. Bone should be decalcified and woods of high density should be softened (see Gray, 1954) before they are embedded in celloidin. Parlodion is a nonexplosive form of celloidin (nitrocellulose). Beware of old stocks of nitrocellulose that are not specified as nonexplosive. When dried out they are exceedingly dangerous.

DEHYDRATION

Begin dehydration at the same aqueous concentration as the fixative or washing solution.

Plant Tissues

1. Finish dehydration of fixed tissues in the following series for 1 hour (root tip) to 8 hours (twig) each: 85, 95, 100 percent ethanol.
2. Continue dehydration in 100 percent ethanol, change at least three times in 24 hours.

Animal Tissues

1. Dehydrate fixed tissues in the standard series, 1 to 2 hours each: 35, 50, 70 percent ethanol.
2. 95 percent ethanol, 2 hours.
3. 100 percent ethanol, two times, 4 hours each.

ORIENTATION OF TWIG SECTIONS

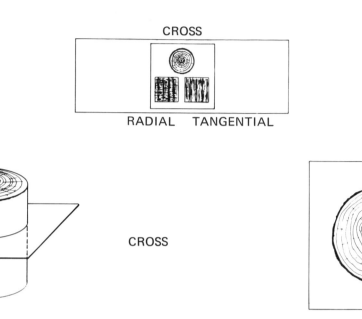

CROSS

RADIAL   TANGENTIAL

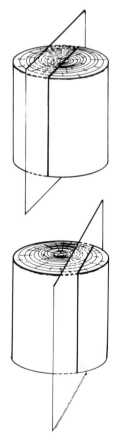

CROSS

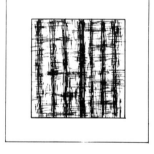

RADIAL

TANGENTIAL

Figure 2–3.   Placement of twig sections.

INFILTRATION

1. Ether-alcohol solvent (anhydrous diethyl ether : absolute ethanol, 1:1); 4 (animal tissue) to 24 hours (plant tissue). Parlodion dissolves slowly so make up the solutions several days in advance. The time schedule for infiltration lists minimal times; each student must adjust his schedule to suit his needs.
2. Infiltrate in a jar with a tight screw-cap at room temperature *in the dark.* Two percent Parlodion in ether-alcohol (2 gm in 100 cc), 24 hours (animal) to 1 week (plant).
3. Four percent Parlodion in ether-alcohol (4 gm in 100 cc), 24 hours (animal) to 1 week (plant).
4. Eight percent Parlodion in a covered Stender dish in the dark.

Check it every day. Add more 8 percent Parlodion as necessary to keep the dish three-quarters full; swirl the dish to mix the contents thoroughly. As the Parlodion thickens, be sure that the tissues are spaced in the dish so they will not interfere with each other when you trim them. Do not allow the Parlodion to thicken too fast or it will have bubbles and an uneven consistency. The tissue is ready to trim and mount on a fiber block when the Parlodion medium feels dry like hard rubber and does not leave a fingerprint.

(Do not pour waste solutions of Parlodion into the sink: Pour them into a waste jar. Waste solutions can be evaporated down and the recovered Parlodion can be used again.)

BLOCKING

Soak a fiber block in ether-alcohol for 10 minutes. Wrap a collar of firm, white paper around the block so that it forms a box slightly deeper than the length of the tissue block. Secure the collar with thread, not a rubber band. Moisten the fiber block and paper with ether-alcohol. Cover the corrugated face of the fiber block with 8 percent Parlodion.

Cut the tissue out of the Parlodion matrix in the Stender dish with a razor blade or scalpel. Trim the tissue block to cutting size and moisten the end opposite the cutting face with ether-alcohol. Place it, moistened end down, in the Parlodion covering the face of the fiber block. Fill the paper collar with 8 percent Parlodion so that it covers the tissue block. Let the Parlodion stand in the air for a few minutes to rid it of air bubbles, then place it in a jar with about 1/2 inch of chloroform in the bottom. Cover the jar tightly and leave it for 20 minutes. Then add enough chloroform to cover the whole block, cover the jar tightly once more, and leave it overnight. Blocks may be stored in 70 percent alcohol.

G. PETERFI'S CELLOIDIN PREIMPREGNATION TECHNIQUE

(Pantin, 1948)

The advantages of Peterfi's method include the support of delicate or brittle objects with celloidin, paraffin embedding for serial sectioning, and an ease and speed of execution not found in double-embedding techniques. In addition, not enough celloidin is present in the sections to interfere with staining.

1. Dehydrate fixed tissues to 100 percent ethanol.
2. One percent celloidin (Parlodion) in methyl benzoate (1 gm in 100 cc); 3 to 5 hours. Run the methyl benzoate-celloidin mixture to the bottom of the jar by sliding it down the sides. The tissue in ethanol will rise up on the ethanol-benzoate interface. The tissue will sink into the methyl benzoate in about 3 hours.
3. Fresh 1 percent celloidin in methyl benzoate, 3 hours.
4. Fresh 1 percent celloidin in methyl benzoate, 12 to 24 hours. Specimens may be stored indefinitely.
5. Benzene (not xylene), 15 minutes.
6. Fresh benzene, 15 minutes.
7. Benzene that has previously been saturated with paraffin on top of the oven at about 30°C, 30 minutes.
8. Paraffin in the 56°C oven as usual. Not more than 4 hours.

H. GELATIN EMBEDDING FOR FROZEN SECTIONS

Gelatin is a good embedding medium when frozen sections are required of small fragile specimens such as *Hydra sp.,* insect tissue, and samples with a high lipid content when time is not an important factor. Generally specimens are embedded in gelatin, hardened in 10 percent formalin, and then frozen and sectioned on a cryostat microtome or a freezing microtome.

FIXATION

Fix small pieces of tissue in Baker's formal-calcium fluid for 24 hours or more. Wash them in running water overnight.

INFILTRATION

1. 10 percent gelatin, 8 to 12 hours at 37°C.

   Knox gelatin 10 gm
   Distilled water 100 cc
   Phenol crystals 0.5 gm
   (or 1:1 dilution of 20 percent gelatin)

2. 20 percent gelatin, 12 hours to overnight at 37°C.

   Knox gelatin 20 gm
   Distilled water 100 cc
   Phenol crystals 1 gm

3. Pour 1/8 inch of 20 percent gelatin (keep it liquid on a warming table) into a paper boat. Allow it to harden at room temperature (a few minutes).
4. Fill the boat with molten 20 percent gelatin and orient the tissue with forceps warmed over an alcohol lamp.
5. Carefully place the material in a refrigerator or cold room until the gelatin is thoroughly set.
6. Trim the gelatin block close to the specimen (within 1 mm).
7. Harden by immersing in formal-calcium fluid overnight.
8. Cut sections with a cryostat microtome or freezing microtome (Chapter 3, Section C). Some technicians wash blocks several hours in running water before they section them.

Chapter 3

MICROTOMES

A. ROTARY MICROTOME FOR PARAFFIN SECTIONS

BASIC EQUIPMENT

Most rotary microtomes work in a similar manner. The tissue, held tightly in a chuck, is moved across the edge of a stationary knife. The advance is controlled by a wedge moving across an inclined plate or by a threaded shaft. Thickness control is usually in increments of 1 micron, starting with 1 micron. Every microtome must be kept well oiled and cleaned. Any wear of parts will loosen the action and cause faulty sectioning.

In addition to the microtome, you need a sharp knife. Sharpen the microtome blade frequently to maintain a sharp edge. A badly worn blade is difficult to recondition. Some laboratories issue razor blades and special holders to students. Gem blades are very good for this purpose.

Additional equipment needed: a small paint brush, forceps, several flat, cardboard boxes (ladies' nylon stocking boxes are the right size), a small bottle of chloroform, and a 1/2-inch paint brush for sweeping away discarded ribbons.

BASIC DIRECTIONS FOR THE ROTARY MICROTOME

1. The microtome should be clean when you start. Brush away all discarded paraffin ribbons. Test the action of the hand wheel. If it is stiff, oil it and clean out the interior of the mechanism. If the block holder is advanced toward its maximum, retract it fully and reset the mechanism.
2. Lock the hand wheel or leave in the position at which the block holder is in its highest position.

3. Clamp the specimen into the chuck of the block holder. Adjust the square face of the paraffin block vertically, horizontally, and with its upper and lower edges parallel with each other and parallel with the base of the microtome. The care with which the block is initially adjusted is important to the success of sectioning.

4. Wipe the microtome blade carefully with chloroform to clean the edge. Place the microtome blade in the blade holder (or razor blade holder with its clamped razor blade). Carefully check the angle of the blade, the firmness of its seating in the holder, and its clearance of the paraffin block and block holder. The angle of the center line of the blade (Fig. 3–1) with the face of the block should be about 20°.

5. If necessary, readjust the paraffin block (see number 3). Some very careful trimming can be done here with a razor blade. Do not allow the razor blade to come in contact with the edge of the microtome blade. Watch your fingers—the blade is sharp!

6. The upper and lower edges of the block must be parallel to each other and to the edge of the knife.

7. Unlock the hand wheel and lower the paraffin block until its face is level with the edge of the microtome blade. Unlock the blade and bring it up to but not in contact with the face of the block. Relock the microtome blade firmly in position.

8. Set the thickness scale as desired. Students should start at 10 microns. When you use a microtome that has ratchet wheels for determining the thickness of section (e.g., A.O. 820), never set the thickness setting at an intermediate position. Always set it at a full (click) stop to avoid damaging the ratchet teeth.

9. Relax and turn the hand wheel with a steady rhythm. Allow the first inch or two of paraffin ribbon to move down the blade until the ribbon begins to bow up near the edge of the blade. Slip your forceps or a dissecting needle under the ribbon at this point and lift the ribbon, still attached to the edge, clear of the knife surface. Hold the ribbon up but do not put any pull on its attachment with the blade, thereby separating the ribbon from the cutting edge or making a weak joint with the next section to be cut. This will take a little practice. Keep turning the hand wheel with a steady rhythm.

10. When the ribbon is about 8 inches long, detach it from the edge of the blade with an *upward* stroke of your paint brush. Even the bristles of the paint brush will nick the edge of the blade, so always brush *away* from the cutting edge of the blade. Lay the ribbon, dull surface up, in your ribbon box and cut another strip.

11. When you are through sectioning, or when you leave the microtome for any reason, you must remove the microtome blade and put it in its box. Many laboratory accidents result from carelessness with the microtome blade.

SECTIONING PROBLEMS AND POSSIBLE REMEDIES
(Modified from Richards, 1959)

1. If the ribbon is crooked rather than straight:

   a. The upper and lower edges of the block face are not parallel to each other or to the knife edge.
   b. The knife edge may be uneven. Try another section of the knife edge.

2. If a ribbon will not form and each section comes off separately:

   a. The blade edge is dull. Sharpen it.
   b. The knife has the wrong tilt. Adjust it.
   c. The upper and lower edges of the paraffin block face are crumbled or rounded. Retrim with a sharp razor blade.

3. If the sections are compressed or folded:

   a. The knife is dull. Sharpen it.
   b. The knife angle is set too close to the vertical. Increase the angle.
   c. The knife is clogged up with paraffin. Clean the edge with chloroform.
   d. The sections are too thin. Increase the thickness setting.
   e. The paraffin is too soft because the room is warm. Pack the blade with ice cubes. Sometimes holding an ice cube against the face of the block helps.
   f. The paraffin is still contaminated with xylene. Re-embed.

4. If the specimen crumbles or falls out of the paraffin section:

   a. The tissue is inadequately dehydrated or cleared. Dissolve off the paraffin, redehydrate, and clear.
   b. The tissue is too hard and compact (e.g., liver). Soak it in water or 70 percent ethanol to soften it.
   c. The tissue was in the paraffin oven too long at too high a heat. Throw it away.
   d. The tissue is too hard for a paraffin matrix (e.g., bone, cuticle, heavy plant fibers, or xylem). Try celloidin embedding.

5. If the ribbon continuously splits or scratches vertically:

   a. Crystals or dirt particles are caught on the knife edge or on the paraffin face. Wipe the knife edge carefully and clean it with chloroform.
   b. The knife edge is nicked. Use another section of the knife or sharpen it.

6. If the ribbon wraps itself around your finger, the knife blade, or anything else:

   a. The problem is static electricity, which is often present in very dry rooms. Let boiling water stand nearby, section in the morning hours, or make short ribbons.

B. SLIDING MICROTOME FOR CELLOIDIN SECTIONS

Celloidin embedding is used for large or very hard specimens. As a result, the microtome for celloidin sections is generally quite large and has a heavy base. The sliding microtome is the instrument most commonly used to prepare celloidin sections. The block is held stationary while a heavy blade is moved across its face at an angle. Beware of the moving blade, as it can cause serious injury. One manufacturer provides a heavy-duty rotary microtome with the blade clamped in a horizontal position and a block holder which moves horizontally just under the blade edge.

Celloidin blocks are stored in 70 percent ethanol and must never be allowed to dry. The ethanol used in cutting and the sliding action of the guide surfaces of the cutting mechanism can cause corrosion and rapid wear to the microtome. Therefore, extreme care must be taken in cleaning and oiling all the sliding surfaces. While cutting, the knife holder should float on a film of oil. After use, wipe, dry, and re-apply oil to all surfaces.

Basic equipment needed: a small paint brush, forceps, a sharp knife, a beaker of 70 percent ethanol, several Stender dishes with 70 percent ethanol.

BASIC DIRECTIONS FOR THE SLIDING MICROTOME

1. Clamp the fiber block on which the celloidin block is mounted into the block holder. Leave a few millimeters of the fiber block exposed above the jaws of the holder so that compression by the jaws will not loosen the celloidin block.
2. Adjust the face of the celloidin block to the horizontal. Lower the advance mechanism so the block will not be endangered while the knife is being adjusted. Always set the block first before setting the knife in the knife clamp. Keep the block moist with 70 percent ethanol with the paint brush.
3. Push the knife holder back so the knife will be clear of the block. After clamping the knife firmly in the holder, adjust it to a 15° vertical angle. The horizontal angle of the knife depends a great deal on the specimen. For celloidin sections of twigs, start with 30 or 40° angle between the knife edge and the direction of travel of the knife holder. Each operator must discover his own best adjustments. Because the knife is set at a horizontal angle in relation to the block, it will encounter a corner of the block before it cuts the section with a slicing motion. The knife is the movable part of the cutting mechanism of the sliding microtome. Be very careful to prevent it from slipping and damaging the block or your finger.
4. Bring the knife directly over the face of the block. Adjust the block face so it is horizontal and parallel to the edge of the blade. With the manual advance, raise the block face until the ethanol film on the block face makes contact with the knife edge.

5. Set the automatic advance mechanism to 30 microns. Move the knife back and forth with a firm, even motion. After each passage of the knife, moisten the block face and the knife edge with 70 percent ethanol.

6. After the knife has made a full section at 30 microns, adjust the advance mechanism for the desired thickness.

7. Each section must be removed from the knife edge with a paint brush and transferred to a dish of 70 percent ethanol. Then the surface of the block and the edge of the knife must be remoistened with alcohol before the next section is cut.

8. If the sections must be kept in order, intersperse each section in the Stender dish with a disc of paper that can be numbered.

9. When you have finished with the cutting, remove, carefully dry, and put away the knife. Replace the block in the jar of 70 percent ethanol. Carefully wipe down the microtome and re-apply oil to all surfaces.

10. Sections are stained individually before they are mounted on a slide. See the directions for celloidin sections under Delafield's Hematoxylin and Eosin Y (Chapter 4, D).

SECTIONING PROBLEMS AND POSSIBLE REMEDIES
(Modified from Richards, 1959)

1. If the sections are scratched or split:

    a. The knife is nicked. Move the knife edge or sharpen it.
    b. Dust particles are in the celloidin matrix. Filter the Parlodion solution before using it again.

2. If the sections fall out of the celloidin matrix:

    a. The block was insufficiently dehydrated or infiltrated. Dissolve off the celloidin and reprocess.
    b. The celloidin is too soft. Try hardening it in chloroform.

3. If the sections are uneven in thickness:

    a. One of the microtome adjustments is still loose. Tighten all the screws and check the firmness of the celloidin block on the fiber block.
    b. The knife holder is moved with an uneven motion. Relax and move the knife evenly. Let the knife do the cutting.
    c. The knife angle is too low. Increase it.
    d. The celloidin block is drying out. Soak it in 70 percent ethanol and take greater care to keep it moist while you section.
    e. The knife is dull. Sharpen it.

C. MICROTOMES FOR FROZEN SECTIONS

CRYOSTAT MICROTOME

The cryostat microtome was developed for rapid production of thin, unfixed sections to be used for histochemical studies and for examination of biopsies during surgery. Basically it is a rotary microtome enclosed within a refrigerated cabinet for which the optimum temperatures for a large number of tissues can be adjusted.

Good sections depend on a very sharp knife and on the rapid freezing of tissues to an optimum temperature. Many tissues will fracture during sectioning unless they are supported by a matrix. Tissues can be embedded in gelatin or surrounded by gum arabic during the freezing process (see Humason, 1967). For best results, completely immerse the tissue in OCT Compound (Lab-Tek, Westmont, Illinois) while freezing. The freezing process must be very rapid to prevent the formation of large ice crystals. The quick-freeze stages of commercially available cryostats must all be supplemented with applications of chips of dry ice, short bursts of carbon dioxide gas, or Freon from a pressure can.

Additional equipment needed: A paint brush and two pairs of forceps (one to be kept in the cabinet). Because the staining procedure is very rapid, have all staining solutions made up ahead of time in Columbia dishes for coverslips or Coplin jars for slides.

BASIC DIRECTIONS FOR CRYOSTAT MICROTOME

1. Set the temperature in the cabinet for $-20°C$. Place the specimen holders, extra forceps, and the knife box with its blade inside to precool.
2. Unfixed tissues are mounted directly on the specimen holders. Lift out a specimen holder and warm its face against your hand. Work with dry hands so you do not stick to chilled equipment. Drop a thin layer of OCT Compound of the proper temperature range (liver: 0 to $-15°C$; heart: -15 to $-30°C$) on the face. Embed the tissue in the viscous compound. Add more compound until the tissue is covered. Cool the holder with the rapid-freeze attachment so the tissue freezes rapidly and evenly or use short bursts from a pressure can of Freon aimed just above the top of the tissue.
3. Place the precooled knife in the knife holder. Adjust the angle to about $10°$. Sections thinner than 10 microns can be cut if chunks of dry ice are taped (with Scotch tape) to the knife to chill it further.
4. Adjust the angle of the specimen holder face. You will generally need a smaller angle than that on the paraffin microtome. Be certain that the face is parallel to the knife so that the metal holder will not hit the knife edge and chip it. If an antiroll device is present, it must be adjusted with

care. If it is absent, each section will have to be guided by a paint brush to prevent curling.

5. Turn the microtome handle with a sharp movement on the downstroke.
6. Remove each section by lifting the antiroll device. Touch the section with the flat surface of a coverslip. The section will immediately thaw and adhere to the coverslip and can be placed in a fixative. Some tissues, such as muscle, that do not adhere easily to the coverslip should be dried for 30 seconds at room temperature before they are fixed.
7. See the directions for staining frozen sections under Delafield's Hematoxylin and Eosin Y (Chapter 4, D).

SECTIONING PROBLEMS AND POSSIBLE REMEDIES (Ames Lab-Tek, 1965)

1. If the sections collapse on the edge of the blade:
   a. The tissue and/or the knife are not cold enough. Wait until they cool enough in the closed cabinet.
   b. The knife is dull. Sharpen it.
2. If the sections crumble:
   a. The sections are too cold. Readjust the cabinet temperature.
3. If the sections tear or vary in thickness:
   a. The knife is dull. Sharpen it.
   b. The microtome mechanism is worn. Have the machine serviced.
   c. The knife is dirty. Clean it.
   d. The tissue is not cold enough. Check the thermostat.

FREEZING MICROTOME

The clinical freezing microtome is a simple microtome that can be clamped to a bench and is fitted with a carbon dioxide freezing attachment. It is essentially a sliding microtome. The blade moves in an arc across the tissue which is frozen on a stationary freezing stage. The advance mechanism is either automatic or manual and raises the freezing stage after each pass of the knife. Sliding microtomes, of the celloidin type, also may be equipped with freezing attachments and make excellent freezing microtomes.

The freezing attachment usually is a hollow metal chamber connected with a cylinder of carbon dioxide gas. Cooling is accomplished by carefully releasing the gas in short bursts into the chamber. Expansion of the gas cools the chamber rapidly and freezes the specimen resting on top of the chamber. Several thermoelectric freezing stages (which depend on the Peltier effect) are now available. Both cooling and heating can be controlled quite easily. These stages adapt very easily to any of the sliding microtomes.

Additional equipment needed: a small paint brush, forceps, and a dropper bottle of distilled water.

BASIC DIRECTIONS FOR CLINICAL FREEZING MICROTOME

1. Check the microtome to ensure that it is well oiled, all the screws are tight, the parts move freely, and the valve to the carbon dioxide gas cylinder opens and closes easily.
2. Place a small square of filter paper (slightly larger than the block) on top of the freezing stage. Saturate it with a drop of water.
3. Place the tissue embedded in a gelatin block (Chapter 2,H) on the filter paper and add another drop of water. The water must form a support at the base of the block but must not build up around the portions of the block to be cut. Ice will interfere with the knife and cause uneven sections.
4. Adjust the height of the block until it is about a fourth of an inch beneath the path of the knife. Release the carbon dioxide gas in short spurts to freeze the tissue. Hold the block down on the freezing stage with forceps until the base begins to freeze.
5. As the block freezes, swing the knife over the block so that the gas will be deflected onto the top of the block and freezing will be speeded. This will also cool the knife.
6. With the advance mechanism, raise the block so that the knife can begin making 15 micron sections. Experience will determine the best cutting temperature. The sections will crumble if the block is too cold. The sections will dissolve into a smear if the block is too soft. The routine should consist of alternately cooling the block with bursts of gas and rapidly cutting a number of sections when the block has warmed to the correct temperature.
7. Remove the section or sections with the small paint brush or the tip of the little finger. Place the sections in a dish of distilled water.
8. Sections of gelatin-embedded tissue are easily handled with a small glass rod bent in the shape of a golf stick. The sections are stained (see Chapter 4,L for Sudan Black B) and mounted in glycerine-jelly.

SECTIONING PROBLEMS AND POSSIBLE REMEDIES
(Modified from Richards, 1959)

1. If the sections crumble:
   a. The block is not hard enough. Speed the freezing time by releasing gas in longer spurts.
   b. The knife is dull. Sharpen it.
   c. The knife is warm. Cool it. Strap dry ice to it with Scotch tape if desired.

2. If the sections vary in thickness:

    a. The knife is dull. Sharpen it.

    b. Some of the microtome screws or adjustments are loose. Tighten all the adjustable screws.

    c. The microtome is worn. Have it serviced.

    d. The block has separated from the freezing stage. Thaw and refreeze.

D. THE MICROTOME KNIFE

Modern microtome knives are made of high-grade steel soft enough to be ground to a very sharp edge and hard enough to hold the sharp edge during contact with paraffin and embedded tissue. Microtome knives are not made of stainless steel and care must be taken to prevent any rusting. A rusted edge on a knife is a pitted and, therefore, a useless, edge. The knives used with the freezing microtome, naturally, are the ones most frequently damaged by rust.

Microtome knives are generally equipped with a "back," which is a tubular piece of steel that slides over the back of the knife and determines the angle at which the knife rests when it lies on the sharpening stone or plate.

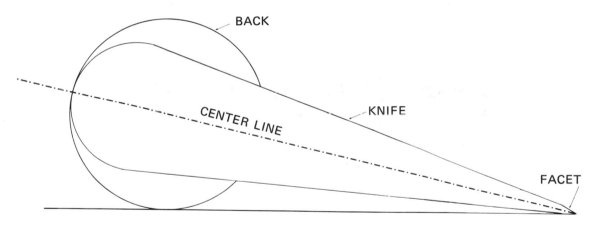

Figure 3-1. Diagrammatic cross section of a microtome blade.

The angle at which the knife is sharpened determines the angle between the facets of the edge and, therefore, its cutting qualities. Every knife that is to be hand sharpened must be equipped with its own, individual back. Knives that are to be machine sharpened do not need a back because each machine is set at a constant angle. Do not sharpen a microtome knife both by machine and by hand at different times because the knife angles will always be slightly different. Never test a knife by cutting a hair or a thread. You will only dull it and have to sharpen it again.

Razor blades make fairly good knives for the beginner or for technicians who have no adequate sharpening equipment. The razor blades are clamped in a special knife clamp and can be discarded when nicked or dull. For best results use a rather thick razor blade clamped in the holder close enough to the cutting edge to prevent vibration but far enough away to prevent hitting the block with the holder. Commercial blades are available to be used with the razor-blade holder. A Gem double-edged blade that has been split in half lengthwise gives satisfactory results.

The cutting edge of a knife can be inspected through the light microscope. Place the knife on a wooden block that will hold it at a tilt of 20° with the cutting edge uppermost. Remove the stage clips or mechanical stage and carefully set the wood block on the microscope stage so that the cutting edge is just under the 10X objective. Do not hit the cutting edge against the objective. Illuminate the knife from an angle above the cutting edge with a bright light, such as a dissecting microscope illuminator. The edge of a sharp knife will be seen as a straight, fine line of reflected light. A dull or nicked knife will reflect more light and will be seen as a serrated line with interruptions identifying the nicks in the edge.

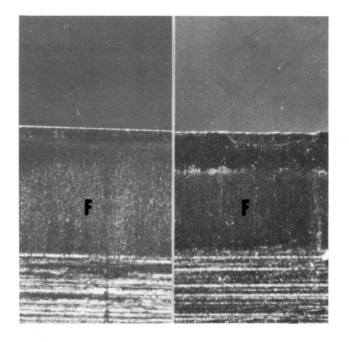

Figure 3-2.   Facet (F) and edge of a sharp (left) and a dull (right) microtome
blade, shown under reflected light.

Chapter 4

STAINING

A. STAINS

Unless they are stained, cell organelles lack enough contrast to be easily distinguished by the human eye. Like primary fixatives, no individual staining solution can adequately provide all the information needed about a cell or group of cells. For general work, two stains of contrasting colors (stain and counterstain) are usually selected to differentiate between nucleus and cytoplasm. Others, used singly or in combinations, may provide specific information concerning the chemical and structural characteristics of all organelles.

You should understand as much as possible about the reaction characteristics of each stain before you use it. The action of many stains is unknown, but a few categories can be distinguished. Many stains have an auxochrome (reactive portion of the molecule that unites with tissue elements) and a chromophore (portion of molecule that imparts its characteristic color).

The auxochrome may bind directly, by salt formation, with a particular portion of the cell, by absorption, or by both means. Some stains, such as hematoxylin (which alone is a very weak dye), use an intermediate binding agent, or mordant, to form a bond with the tissues. Stains such as Sudan Black B are dissolved directly into parts of the cell. Metals, such as silver in Holmes' method for nerves, can be precipitated onto certain structures to differentiate them from other parts of the cell. The main purpose of staining is to provide a specific part of a cell with color contrast that will not diffuse away or change its color during the period of examination.

B. GENERAL STAINING ROUTINE

Use clean glassware. Slides can be conveniently stained in Coplin jars. However, unmounted sections, frozen or embedded in celloidin, are floated in Stender dishes. Set up jars with solutions before you start. Label each jar clearly and keep them covered at all times. Wipe or drain slides with absorbent tissue between each solution. If you must inspect the wet section for stain quality, protect the microscope by placing a lantern-slide glass on the stage underneath the wet slide. Beware of

shortcuts. Always keep the solutions fresh, especially the alcohol series for hydra-
tion and dehydration. The final dehydration before applying the coverslip is critical.
If any water is present, the xylene will cloud and the stains will diffuse or fade.
Never let slides dry at any time until the coverslip is on.

HYDRATION AND DEHYDRATION

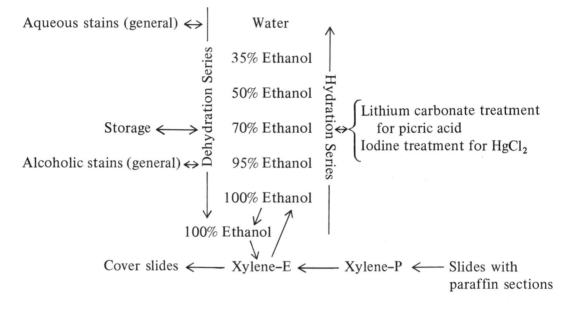

Slides should be immersed in xylene for 5 minutes and in each alcohol solution for
1 to 5 minutes each. To remove any mercury deposit add enough iodine solution
(Chapter 2, B) to 70 percent ethanol to make it dark brown and treat the slides for
about 5 minutes. Picric acid will block the stain action. If any yellow remains in the
sections after Bouin's fixative, treat them in 70 percent ethanol saturated with lithium
carbonate until the sections show white. If the sections show a tendency to float off
the slide, coat the slide and sections with a thin film of Parlodion. After 100 percent
ethanol in the hydration series, dip the slide into 1 percent Parlodion in ether-alcohol,
lift it out and let it drain, then place it directly in 70 percent ethanol.

COVERING

After the slide has been dehydrated and cleared in xylene, remove it and drain off
the excess xylene. *Do not allow the sections to dry.* Check to make sure your sections are on the upper surface of the slide. This is easily done by feeling the engraved
identification on the undersurface at the end of the slide. Place a small drop of
Permount on the middle section. Carefully lower a coverslip at an angle over the
section and set the covered slide aside to dry for 5 days. If it is placed flat in an oven
at 45°C it will be ready for gentle cleaning in 24 to 48 hours.

CLEANING AND LABELING

Scrape the excess hardened Permount away from the slide and coverslip with a razor blade. Be gentle because the Permount under the coverslip may still be soft. Wipe the slide gently with a tissue dampened with xylene. Do not jar the coverslip. Then wipe the slide with an alcohol-dampened tissue. Just before attaching the slide label to the left end of the slide, dip that part of the slide into 50 percent alcohol and polish it well. If the slide is dirty, slide labels will soon become detached.

The slide label must contain the following information (in black ink):

> Tissue (orientation, if any)
> Organism
> Fixative (thickness of section)
> Stain
> Date (student's initials)

C. METHYLENE BLUE–PHLOXINE

Methylene blue–phloxine is a simple method of distinguishing between the basic and acidic reacting portions of the cell. It is especially effective in demonstrating the effects that different fixatives have on various tissues.

FIXATIVE

Any general fixative is adequate.

PROCEDURE

1. Xylene (labeled Xylene–P), 5 minutes.
2. Xylene (labeled Xylene–E), 2 minutes.
3. 100 percent ethanol, 2 minutes.
4. 95 percent ethanol, 2 minutes.
5. 70 percent ethanol, 2 minutes.
6. If the tissues were fixed in Helly's fixative, remove the mercury deposit in 70 percent ethanol and enough iodine solution (see Helly's fixative) to make the solution dark brown, 5 minutes. If the tissues were fixed in Bouin's fixative, leave the slides in 70 percent ethanol with a small amount of lithium carbonate until *all* the yellow color has disappeared.
7. 70 percent ethanol, 2 minutes.
8. 50 percent ethanol, 2 minutes.

9. 35 percent ethanol, 2 minutes.
10. Water, 2 minutes.
11. Phloxine solution, 1 minute.
12. Rinse well in distilled water.
13. Methylene blue solution, 1 minute.
14. Dip the slides once or twice in 0.2 percent aqueous acetic acid to wash off any excess blue stain.
15. 95 percent ethanol, 1 minute. Dip the slides up and down several times.
16. 95 percent ethanol, 1 minute. Dip the slides up and down several times.
17. 95 percent ethanol, 1 minute. The third ethanol solution should be colorless. Discard each solution as it becomes distinctly blue.
18. 100 percent ethanol, 2 minutes.
19. 100 percent ethanol, 2 minutes.
20. Xylene–E, 5 minutes.
21. Drain off the xylene, add a small drop of Permount, and cover the sections carefully with a coverslip. *Do not allow the slides to dry.*

RESULTS

Blue: nuclei, cytoplasmic basophilia, goblet cell mucin.
Pink: cytoplasm, collagen, muscle fibrils.

OPTIONAL

The basophilic reactions resulting from DNA and RNA can be identified by differential extraction with perchloric acid and comparison of the normal slide with the extracted slide. Three slides of each tissue are needed: one each for the normal stain, the stain after extraction of RNA, and the stain after extraction of both DNA and RNA.

To Extract RNA

After the first hydration of the slides to water, place one of them in 10 percent perchloric acid at 4°C for 12 hours. Neutralize the slide in 1 percent sodium carbonate for 5 minutes at room temperature. Wash it in tap water and stain it by the regular procedure.

To Extract DNA and RNA

After the first hydration of slides to water, place one of them in 5 percent perchloric acid at 60°C for 30 minutes. Neutralize, wash, and stain it in the same way you did for RNA extraction.

The blue stain in the cytoplasm of the normal slide (without extraction) is caused by ribosomal RNA. The blue stain in the nucleus is caused by the acid reaction of DNA. The slide pretreated with perchloric acid at 4°C should have lost its cytoplasmic nucleic acid. Therefore, the cytoplasm will have lost its blue color (stain pink), whereas the nucleus will still contain blue chromatin. The slide pretreated with perchloric acid at 60°C should have lost both cytoplasmic as well as nuclear nucleic acids. Therefore, both the cytoplasm and nucleus will stain pink.

SOLUTIONS

Phloxine Solution

Phloxine B, C.I. 45410	0.5 gm
Distilled water	100 cc
Glacial acetic acid	0.2 cc

A slight precipitate will form. Filter before use.

Methylene Blue Solution

Methylene blue, C.I. 52015	0.25 gm
Azure B, C.I. 52010	0.25 gm
Borax	0.25 gm
Distilled water	100 cc

D. DELAFIELD'S HEMATOXYLIN AND EOSIN Y FOR PARAFFIN, CELLOIDIN, AND FROZEN SECTIONS

Hematoxylin is probably the most universal stain used in histology. Hematoxylin itself (or hematein, its oxidation product) is a rather weak stain and requires a mordant to increase its attraction for tissues. A mordant is a high-molecular weight metal salt that forms a complex both with tissue and with the dye and acts as a strong bonding agent. Delafield's hematoxylin has the mordant, ammonium alum, mixed in with the dye solution and therefore stains in a single step. It is important that all students of microtechnique master this staining method.

FIXATIVE

Any general fixative is adequate.

PROCEDURE FOR PARAFFIN SECTIONS

1. Xylene (labeled Xylene–P), 5 minutes.
2. Xylene (labeled Xylene–E), 2 minutes.
3. 100 percent ethanol, 2 minutes.
4. 95 percent ethanol, 2 minutes.
5. 70 percent ethanol, 2 minutes.
6. After using Helly's fixative, remove the mercury deposit in 70 percent ethanol and enough iodine solution (see Helly's fixative) to make the solution dark brown, 5 minutes. After using Bouin's fixative, leave the sections in 70 percent ethanol with excess lithium carbonate until all the yellow color has disappeared.
7. 70 percent ethanol, 2 minutes.
8. 50 percent ethanol, 2 minutes.
9. 35 percent ethanol, 2 minutes.
10. Water, 2 minutes.
11. Delafield's hematoxylin, approximately 5 minutes.
12. Rinse off the stain in running tap water and check the slide on a protective glass-plate under the microscope. Nuclei should be dark blue and cytoplasm gray to very pale blue.
13. 35 percent ethanol, 2 minutes.
14. 50 percent ethanol, 2 minutes.

If the cytoplasm is blue and the sections are overstained:

    a. Acid alcohol (two drops of concentrated hydrochloric acid in a Coplin jar of 70 percent ethanol) until the sections are red-purple. Be careful not to remove too much stain.
    b. 70 percent ethanol, 1 minute.

15. 70 percent ethanol saturated with lithium carbonate, 5 minutes or until all the cells show a blue color. If the cells are still overstained, repeat the acid-alcohol treatment. If the cells are now understained, hydrate back down the alcohol series and restain.
16. 95 percent ethanol, 2 minutes.
17. Eosin Y solution, 2 to 5 minutes.
18. 95 percent ethanol, 2 or 3 minutes to rinse off the eosin stain. If the eosin is too dark, leave the slide in 95 percent ethanol to destain slowly.
19. 100 percent ethanol, 2 minutes.
20. 100 percent ethanol, 2 minutes.
21. Xylene–E, 5 minutes.
22. Drain off the xylene, add a small drop of Permount, and cover the sections carefully with a coverslip. *Do not allow the slide to dry*.

RESULTS

Blue: nuclei.
Pink: Cytoplasm, muscle fibrils, connective tissue fibers.

SOLUTIONS

Delafield's Hematoxylin

Hematoxylin, C.I. 75290	4 gm
95 percent ethanol	25 cc

Dissolve. Mix into the following saturated solution:

Ammonium alum ($NH_4 Al(SO_4)_2 \cdot 12 H_2O$)	36 gm
Distilled water	400 cc

Let the solution stand a week exposed to light, lightly covered. Filter. Add

Glycerine	100 cc
Methanol, 100 percent	100 cc

Let the solution stand 6 weeks to ripen. The stock will keep indefinitely.

Eosin Y Solution for Paraffin and Frozen Sections

Eosin Y, C.I. 45380	0.5 gm
95 percent ethanol	100 cc

Eosin Y Solution for Celloidin Sections

Eosin Y, C.I. 45380	1 gm
Distilled water	100 cc

PROCEDURE FOR CELLOIDIN SECTIONS

Float the sections individually in solutions in Stender dishes. Transfer the sections by means of a paint brush or a section lifter made by bending a glass rod into the shape of a golf club.

1. 70 percent ethanol, sections are cut and stored.
2. 50 percent ethanol, 30 seconds.
3. 35 percent ethanol, 30 seconds.
4. Delafield's hematoxylin, 5 to 30 minutes.
5. Tap water; change the water several times until the sections are blue. Any excess celloidin should be trimmed away with a sharp razor blade.
6. Acid alcohol (70 percent ethanol and hydrochloric acid); 2 to 3 seconds. The celloidin should be decolorized.

7. Tap water; saturated with lithium carbonate until all the cells are blue.
8. 1 percent aqueous Eosin Y, 2 minutes.
9. Tap water; about 5 minutes or until the eosin is properly differentiated.
10. 70 percent ethanol, 2 minutes.
11. 95 percent ethanol, 2 minutes.
12. Dip the tip of a clean slide into alcohol and slide the section up onto it with a paint brush.
13. Flood the slide with 100 percent ethanol, 2 times, 1 minute each.
14. *In the hood,* flood the slide with carbol-xylene (saturated phenol in xylene) and drain. Repeat until the sections are clear. This solution is very caustic. Do not allow it to touch your skin or eyes.
15. Flood the slide with xylene, drain, and wipe off the excess fluid. Repeat the procedure twice and wipe. It is important to remove all the phenol possible.
16. Cover with a number 2 coverslip. Use fairly thick Permount. Weight the coverslip while the Permount is hardening.

PROCEDURE FOR CRYOSTAT FROZEN SECTIONS

The sections are picked up directly on very clean slides or coverslips and will thaw out as soon as they touch the warm slide.

1. Acetic alcohol (glacial acetic acid : 100 percent ethanol, 1:3), 15 seconds.
2. Distilled water, 15 seconds.
3. Delafield's hematoxylin, 15 seconds.
4. Distilled water, 15 seconds.
5. Tap water, until blue.
6. Eosin in 95 percent ethanol, 5 seconds.
7. 95 percent ethanol, rinse off the excess eosin.
8. 100 percent ethanol saturated with lithium carbonate, 2 times, 15 seconds each.
9. Xylene, 30 seconds or until the section is clear.
10. Mount with Permount.

E. HEIDENHAIN'S IRON HEMATOXYLIN

Heidenhain's iron hematoxylin is a two-step hematoxylin method. The tissues are exposed to the mordant first and then are placed in the stain solution. This method creates sharp contrasts between parts of the cell and works well with almost any fixation.

FIXATIVE

Any general fixative is adequate; Helly's is preferred.

PROCEDURE

1. Deparaffinize and hydrate the slides to water (treat for mercuric chloride if necessary).
2. Distilled water, 5 minutes.
3. 4 percent iron alum (mordant), 1 hour to overnight.
4. Rinse in distilled water. Change the water once.
5. 1 percent hematoxylin, 1 hour to overnight. Leave in the stain approximately the same length of time as in the mordant.
6. Distilled water, 5 minutes. Change once or twice during this time.
7. 2 percent iron alum (differentiation); after 5 minutes the sections will become dark gray. Watch the sections under a microscope until the cytoplasm is gray and the nuclei are black. Keep the sections covered with alum solution. Timing is important as tissues differentiate at different speeds. Keep a careful watch—most slides will be differentiated within 10 minutes. If the cells differentiate too fast, try a 1 percent iron-alum differentiator.
8. Running tap water, 1 hour. Any iron alum left in sections will cause fading.
9. Dehydrate the slides to 95 percent ethanol.
10. Orange G, 3 minutes.
11. Rinse quickly in 95 percent ethanol.
12. 100 percent ethanol; 2 times, 2 minutes each.
13. Clear and cover (Permount).

RESULTS

Black: nuclei, cytoplasmic granules, centrioles, elastic fibers.
Orange: cytoplasm.

SOLUTIONS

Stock Hematoxylin

Hematoxylin, C.I. 75290	10 gm
95 percent ethanol	100 cc

Stock solution must ripen at least 3 months. Before you use it, dilute 1:9 (stock:distilled water), filter.

Iron Alum (Ferric Ammonium Sulfate, $FeNH_4(SO_4)_2 \cdot 12 H_2O$)

Mordant: 4 percent alum in distilled water (4 gm in 100 cc).

Differentiator: 2 percent alum in distilled water (2 gm in 100 cc).

Use lavender crystals only; brown crystals have decomposed. Filter before use.

Counterstain

Orange G, C.I. 16230	1 gm
95 percent ethanol	100 cc
0.1 N hydrochloric acid	4 cc

F. SAFRANIN AND FAST GREEN

(Jensen, 1962)

Safranin and fast green, a standard method for plant material, distinguishes between the lignified and cellulose portions of the cell wall.

FIXATIVE

FAA or Carnoy's

PROCEDURE

1. Deparaffinize; hydrate the slides to 50 percent ethanol.
2. Safranin, overnight to 24 hours or longer.
3. Rinse the slides in distilled water to remove any excess stain.
4. 70 percent acidified ethanol (2 drops of concentrated hydrochloric acid in a Coplin jar of alcohol) to remove nonspecific stain, 2 dips.
5. 95 percent ethanol, 15 seconds.
6. 100 percent ethanol, 2 times, 15 seconds each.
7. Fast green FCF solution, 1 to 4 minutes.
8. Differentiating solution, 2 times, 10 minutes each.
9. Xylene, 3 times, 15 minutes each.
10. Cover with Permount and coverslip.

RESULTS

Red: chromosomes, nucleoli, lignified cell walls.
Green: cytoplasm, cellulose cell walls.

SOLUTIONS

Safranin

| Safranin 0, C.I. 50240 | 1 gm |
| 95 percent ethanol | 100 cc |

For use, dilute with distilled water (1:1).

Fast Green

Fast green FCF, C.I. 42053	0.5 gm
Clove oil	50 cc
100 percent ethanol	50 cc

Differentiating Solution

Clove oil	50 cc
100 percent ethanol	25 cc
Xylene	25 cc

G. PERIODIC ACID—SCHIFF METHOD FOR GLYCOGEN

Schiff reagent will combine with aldehyde and become a red-staining compound. Many tissue carbohydrates will yield aldehyde groups when they are oxidized by periodic acid that is specific for paired alcohol groups (1, 2-glycol groups). To identify glycogen specifically, run two additional slides as controls. One slide should be treated with Schiff reagent without pretreatment with periodic acid to identify the natural tissue reactive sites. The second slide should be placed in saliva or freshly made 1 percent diastase (1 gm in 100 cc water) for 1 or 2 hours to remove selectively all glycogen. A careful comparison of the three slides will demonstrate the location of any glycogen in the cells.

FIXATIVE

Carnoy's solution

PROCEDURE

1. Deparaffinize, hydrate to water.
2. Periodic acid, 5 minutes.
3. Running tap water, 5 minutes.
4. Schiff reagent, 10 minutes.
5. Sulfite rinse, 3 times, 2 minutes each.
6. Running tap water, 5 minutes.
7. Distilled water, 1 minute.
8. Delafield's hematoxylin, 2 minutes.
9. Distilled water, rinse off the stain.
10. Tap water, until the section is blue.
11. Dehydrate through the alcohol series to 70 percent ethanol. Destain if necessary (see the directions for Delafield's H & E).
12. 70 percent ethanol with lithium carbonate, 2 minutes.
13. Dehydrate, clear and cover (Permount).

RESULTS

Red: glycogen, mucin, cartilage, starch, chitin.
Blue: nucleus.

SOLUTIONS

Schiff's Reagent (Humason, 1967)

Basic fuchsin, C.I. 42500	0.5 gm
Distilled water	85 cc
Sodium metabisulfite ($Na_2 S_2 O_5$)	1.9 gm
1 N−hydrochloric acid	15 cc

Place in a 150-cc flask or bottle. Leave the material overnight (shake the container occasionally). Add 200 mg activated charcoal, shake for 1 minute, and filter. The solution should be white or pale straw yellow. Repeat the charcoal treatment until the solution is the correct color. Store it in a 100-cc bottle in the refrigerator.

Periodic Acid

Periodic acid (HIO_4)	1 gm
Distilled water	100 cc

Sulfite Rinse

Sodium metabisulfite ($Na_2 S_2 O_5$)	0.5 gm
Distilled water	100 cc

Counterstain

Delafield's hematoxylin (see Chapter 4, D)

H. FEULGEN METHOD FOR DNA

(Humason, 1967)

The Feulgen reaction is specific for DNA. The hydrochloric acid hydrolyzes the nucleic acid and leaves the deoxyribose sugar with an aldehyde group to react with the Schiff reagent.

FIXATIVE

Carnoy's preferred (Bouin's not recommended)

PROCEDURE

1. Deparaffinize and hydrate to water.
2. 1 N–hydrochloric acid at 60°C, 6 minutes for Carnoy's, 8 minutes for formalin or Helly's.
3. Distilled water, 30 seconds.
4. Schiff reagent, in the dark, 2 hours.
5. Bleaching solution, 3 times, 2 minutes each. Wipe the slides carefully at each change.
6. Distilled water, 1 minute.
7. Fast green FCF, 5 to 10 seconds.
8. 95 percent ethanol, rinse.
9. Dehydrate, clear, and cover (Permount).

RESULTS

Red: DNA.
Green: cytoplasm.

SOLUTIONS

Schiff Reagent

See Periodic Acid–Schiff Method for Glycogen (Chapter 4, G).

1 N–Hydrochloric Acid

Concentrated hydrochloric acid (37 to 38 percent assay)	8 cc
Distilled water	100 cc

Bleaching Solution

1 N–hydrochloric acid	5 cc
10 percent aqueous sodium metabisulfite ($Na_2S_2O_5$)	5 cc
Distilled water	100 cc

Make up fresh before use.

Fast Green FCF

Fast green FCF, C.I. 42053	0.05 gm
95 percent ethanol	100 cc

I. ALTMANN'S ANILINE-ACID FUCHSIN FOR MITOCHONDRIA

(Conn et al., 1960)

Altmann's aniline-fuchsin method is the classical method for staining mitochondria. The fixation is lengthy but the staining process is very fast. It is important that the sections be 6 microns or less in thickness so that individual mitochondria can be distinguished.

FIXATIVE

Regaud's solution

PROCEDURE

1. Deparaffinize 5 or 6 micron sections, and hydrate to water.
2. Drop some stain onto the slide (about 2 drops); heat the slide gently over an alcohol flame until the stain begins to steam. Place the slide on a paper towel and let it cool, 5 minutes.
3. Drain off the stain; rinse the slide in distilled water, 3 minutes.
4. Methyl green, 15 to 30 seconds. Timing is critical; methyl green stains as well as extracts the acid fuchsin.
5. Drain and wipe; rinse quickly in 95 percent ethanol.
6. Dehydrate, clear, and cover (Permount).

RESULTS

Red: mitochondria.
Green: nuclei.

If methyl green extracts too much acid fuchsin, mordant sections in 3 percent potassium dichromate for 5 minutes before you stain.

SOLUTIONS

Aniline Water

Distilled water	25 cc
Aniline	4 drops

Shake vigorously and filter. Make up the solution fresh each time. Use aniline *in the hood.* The fumes are toxic.

Aniline-Acid Fuchsin

Acid fuchsin, C.I. 42685	1 gm
Aniline water	10 cc

Filter. Let the solution stand 24 hours before you use it. The stain keeps only 1 month.

Counterstain

Methyl green, C.I. 42590	1 gm
Distilled water	100 cc

J. HOLMES' SILVER NITRATE METHOD FOR NERVES

(Holmes, 1947)

Silver impregnation methods have been used traditionally to demonstrate reticulum fibers (connective tissue), Golgi apparatus, and nerve cells and processes. The silver solution does not act as a stain but impregnates the tissue and, when reduced by a reducing solution, forms a black metal deposit that is quite selective. Pyridine, a fat solvent, facilitates the penetration of the silver nitrate of the developer in fine-grain detail. Gold toning serves to substitute part of the dead black silver color and creates a variable colored image with an improved contrast. Do not allow any metal instruments to come into contact with the silver solutions. Coat the tools used to handle the slides with paraffin.

FIXATIVE

Use almost any fixative except those that contain potassium dichromate, chromic acid, or osmium tetroxide. Holmes used saline-formalin, Bouin's, and Carnoy's.

PROCEDURE

1. Deparaffinize and hydrate to water.
2. Tap water, 10 minutes.
3. Distilled water (glass redistilled), 2 times, 5 minutes each.
4. 20 percent silver nitrate, 2 hours, *in the dark.*
5. Distilled water, 3 times, 10 minutes each.
6. Developer, overnight, in 37°C oven in the dark.
7. Reducer, 2 minutes, warmer than 25°C (on top of the oven).
8. Running tap water, 3 minutes.

9.   Distilled water, 1 minute.
10. 0.2 percent gold chloride, 3 minutes or until colorless.
11. Distilled water, 1 minute.
12. 2 percent oxalic acid, 3 to 10 minutes or until the axon is thoroughly black.
13. Distilled water, 1 minute.
14. 5 percent sodium thiosulfate, 5 minutes.
15. Running tap water, 10 minutes.
16. Distilled water, 1 minute.
17. Dehydrate through the ethanol series, clear, and cover.

RESULTS

Black: neurons, especially axons.
Rose-brown: muscle, other cellular elements.

SOLUTIONS

Silver Nitrate Stock

Silver nitrate	20 gm
Distilled water	100 cc

Boric Acid Solution

Boric acid	1.24 gm
Distilled water	100 cc

Borax Solution

Borax	1.9 gm
Distilled water	100 cc

Pyridine Solution *(Keep in the hood)*

Pyridine	10 cc
Distilled water	90 cc

Developer

Boric acid solution	55 cc
Borax solution	45 cc
Distilled water	394 cc
1 percent silver nitrate (dilute stock 1:18)	2 cc
Pyridine solution	5 cc

Keep in the dark. Make up just before use.

Reducer

Hydroquinone	1 gm
Sodium sulfite	10 gm
Distilled water	100 cc

Gold Chloride Stock

Gold chloride	0.2 gm
Distilled water	100 cc

Oxalic Acid

Oxalic acid	2 gm
Distilled water	100 cc

Sodium Thiosulfate

Sodium thiosulfate	5 gm
Distilled water	100 cc

K. HEIDENHAIN'S AZAN

Heidenhain's azan method gives a beautiful stain for connective tissue and for the granules in the pituitary and pancreas. This particular method gives outstanding results both for general vertebrate and invertebrate material. The full schedule, which gives the most brilliant results, is rather lengthy, and shorter times, which give less intense results, are suggested in parentheses (after Culling, 1963).

FIXATIVE

Helly's or Zenker's fluid.

PROCEDURE

1. Deparaffinize and hydrate the slides to water; treat for mercuric chloride at the appropriate step.
2. Mordant in 3 percent potassium dichromate, 1 hour. If azan does not give an intense red stain, mordant overnight.
3. Distilled water, 5 minutes; make several changes.
4. Azocarmine solution, 1 hour at 60°C, allow to cool for 1 hour (1 hour at 60°C; do not cool before proceeding to next step).

5. Rinse in distilled water.
6. Differentiate in an aniline-alcohol solution, about 1 hour (about 20 minutes). Check the slides under the microscope and differentiate until only the nuclei and red blood cells are bright red and the cytoplasm is pale pink.
7. Stop the differentiation in acid alcohol.
8. Rinse the slides in distilled water.
9. 5 percent phosphotungstic acid, 2 hours (45 minutes).
10. Counterstain, 2 hours with fresh stain, 12 to 24 hours with depleted stain (45 minutes). The counterstain becomes depleted rapidly, especially the aniline blue. Replace the stain after every 50 slides.
11. Rinse the slides rapidly in distilled water and check them under the microscope to ensure that the staining is adequate.
12. 95 percent ethanol, 30 seconds.
13. 100 percent ethanol, twice, 3 minutes each.
14. Clear in xylene and cover (Permount).

RESULTS

Red: nuclei, red blood cells.
Orange to orange-red: acidophil granules, muscle fibers.
Blue: collagen, reticular fibers, mucin, basophil granules.
Pink to gray or colorless: chromophobe cytoplasm, nerve axons.

SOLUTIONS

Azocarmine Solution

Azocarmine B, C.I. 50085	0.2 gm
Distilled water	100 cc
Glacial acetic acid	1 cc

Aniline-Alcohol Solution

Aniline	0.1 cc
95 percent ethanol	100 cc

Counterstain

Orange G, C.I. 16230	2 gm
Aniline blue, WS, C.I. 42780	0.5 gm
Glacial acetic acid	8 cc
Distilled water	100 cc

Potassium Dichromate Solution

Potassium dichromate ($K_2Cr_2O_7$)	3 gm
Distilled water	100 cc

Phosphotungstic Acid Solution

Phosphotungstic acid ($P_2O_5 \cdot 24\,WO_3 \cdot H_2O$)	5 gm
Distilled water	100 cc

L. SUDAN BLACK B FOR LIPIDS

Sudan Black B, a fat-soluble stain, is very specific for neutral lipids. The best results are obtained when the stain is dissolved in propylene glycol (0.7 gm/100 cc); however, the use of ethanol as a solvent gives good results. The following directions are for unmounted, frozen sections fixed in neutral formalin and embedded in glycerine jelly (Chap. 2, H).

PROCEDURE

1. Prepare frozen sections (15 microns) and float them in distilled water; two changes of water, 5 minutes each.
2. 50 percent ethanol, 2 minutes.
3. Sudan Black B, 5 minutes. Try to keep the sections from sinking to the bottom of the dish. Keep the cover of the dish closed.
4. 50 percent ethanol, 3 times, 1 second each.
5. Distilled water, 5 minutes.
6. Mount the stained sections in glycerine jelly: (a) With the tip of a scalpel remove a small chip (about 2 mm across) of glycerine jelly from the stock dish. Place the chip in the center of a clean slide and heat the slide gently on the warming table (45°C) to melt the jelly. (b) Or, heat a test tube of glycerine jelly in a beaker of warm water until the jelly melts and place a drop of liquid jelly on a warm slide lying on the warming table. Drain as much water as possible from the stained section and place it in the drop of glycerine jelly. Take a circular coverslip, breathe on the bottom surface, and lower it carefully and squarely onto the section. Remove the slide from the warming table. Allow it to cool and solidify. Semipermanent preparations can be made by sealing the edges of the coverslip with Duco cement, fingernail polish, or any of the commercial sizing cements and gold sizes.

RESULTS

Blue-black to black: lipids, mitochondria, myelin (nerve).

SOLUTIONS

Sudan Black B

Sudan Black B, C.I. 26150	2 gm
70 percent ethanol	100 cc

Make up the solution 3 or 4 days in advance. Filter it twice before you use it. Do not reuse the stain.

Glycerine Jelly

Knox gelatin	8 gm
Distilled water	52 cc
Soak 1 hour and add	
Glycerine	50 cc
Phenol	2 small crystals

Heat the material gently to 60°C, stirring until thoroughly mixed. Keep it in a tightly covered, wide-mouth jar.

Chapter 5

WHOLE MOUNTS

A. GENERAL REMARKS

Many preparations do not require sectioning. Small specimens such as *Hydra,* flat-worms, algae, and some insects, when used as whole mounts, can provide important information. The three-dimensional characteristics of cells in multicellular organisms also can be studied to advantage by macerating or squashing the organism and preserving the isolated cells as whole mounts. This is particularly important to the study of the diploblastic characteristics of *Hydra.*

Temporary slides may be made by covering the specimen in its macerating fluid or in water with a coverslip (preferably round). The edges of the coverslip can be sealed with fingernail polish or Duco cement to prevent rapid evaporation. Such mounts will last several days if they are stored in the refrigerator.

Permanent mounts can be made directly from water by using aqueous mounting media such as glycerol jelly, CMC-10 (Turtox), or Hoyer's medium (Ward's). The edges of the coverslips for these preparations should also be sealed to ensure permanence.

Permanent mounts are best prepared with a standard resinous medium such as Permount. The specimens should be handled carefully on strips of stiff paper because they become very brittle during fixation and dehydration. The mounting resin should be thicker than the standard medium used for paraffin sections. Thick specimens should be mounted in mounting rings or cells or the coverslip should be supported by chips of glass from a broken slide or by short lengths of monofilament fishing line. As the resin dries and shrinks, more medium may have to be added to the edge of the coverslip to keep it fully supported.

B. WRIGHT'S STAIN FOR BLOOD SMEARS

PREPARATION OF BLOOD SMEAR

1. Lay out several clean slides on a clean paper towel. The slides must be clean and free of grease (wash them with soap or Bon Ami, rinse them first in distilled water and then in 95 percent ethanol).
2. Wipe the tip of a finger with alcohol. Dry the finger by holding your arm down at your side and shaking your fingers.
3. Prick the skin of the cleaned finger with a sterile disposable lancet. Throw the lancet away immediately—allow no one else to use it.
4. Wipe away the first drop of blood. Pick up the second drop on the lower edge of one end of a clean slide. Lower the edge to the surface of a second clean slide near its end so that the blood drop makes contact. Push the edge of the upper slide across the surface of the lower slide. An even film of blood one cell thick will be spread across the slide.
5. Wave the smear preparation in the air to expedite its drying.

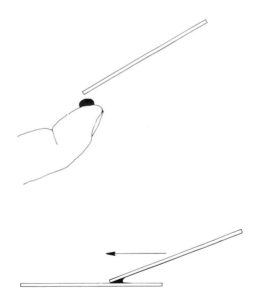

Figure 5-1. Formation of blood film on a slide.

PROCEDURE

1. Flood the smear with Wright's stain (about 10 drops), 1 minute.
2. Add an equal number of drops of buffer solution, 3 minutes.

3. Flush the stain off the slide by dropping distilled water on it and tipping the slide. Rinse the slide in distilled water until the thinner sections of the blood film show pink.

4. Shake off the excess water and wipe off as much of the water as possible without disturbing the smear.

5. Air dry thoroughly. Cover (dilute Permount).

RESULTS

Erythrocytes: pink.
Lymphocytes: dark-purple nuclei, pale-blue cytoplasm.
Monocytes: purple to blue nuclei, pale-blue cytoplasm.
Neutrophils: dark-purple nuclei, lavender cytoplasm and granules.
Eosinophils: purple nuclei, pink granules.
Basophils: purple nuclei, deep-purple granules.

SOLUTIONS

Wright's Stain

Wright's stain	0.1 gm
Neutral methanol, acetone free	60 cc

Grind the materials together thoroughly in a mortar until they are dissolved. Keep them tightly stoppered. Commercial solutions are available.

Buffer Solution (pH 6.5)

2.76 percent monobasic sodium phosphate (NaH_2PO_4 ; 2.76 gm in 100 cc)	68 cc
5.36 percent dibasic sodium phosphate (Na_2HPO_4 ; 5.36 gm in 100 cc)	32 cc

C. MACERATION TECHNIQUE FOR *Hydra*

(Modified from Dyer and Willey, 1969)

Maceration techniques are convenient for the study of the individual cell. *Hydra fusca, Hydra americana,* and *Podocoryne carnea* (a marine hydroid) macerate especially easily. Many stain techniques can be adapted to this method. *Hydra* can be obtained from a commercial biological supply house or can be collected in local ponds.

MACERATION

1. Using a pipette, place a single *Hydra* on a clean glass slide in a small drop of water.
2. Allow it to relax and elongate. The tentacles should be extended fully.
3. When the body and tentacles are extended fully, drop 3 to 5 drops of Haller's fluid directly onto the animal, 3 minutes.
4. Blot away the Haller's fluid and, at the same time, drop distilled water over the *Hydra* to rinse away the acid. The specimen is *very* fragile at this stage.
5. Drop a clean coverslip onto the specimen. The *Hydra* cells will spread out under the coverslip. Tap the coverslip repeatedly but gently with a glass rod to ensure maximum dissociation. Blot away any excess fluid at the edge of the coverslip.
6. Transfer the slide to the top of a piece of dry ice in an insulated container. The slide should freeze immediately.
7. After 1 minute, while the slide is still in the container, slide the edge of a razor blade under the corner of the coverslip and twist it off. Work rapidly. The slide must still be frozen when the coverslip comes off. Thaw the preparation by holding your hand against bottom of slide.
8. Immediately place the slide in a Coplin jar with Carnoy's fluid, 1 minute.
9. 95 percent ethanol, 1 minute.
10. Distilled water, 1 minute.
11. Phloxine, 1 minute.
12. Rinse in distilled water.
13. Methylene blue solution, 1 minute.
14. 0.2 percent acetic acid (2 drops of glacial acetic acid in a Coplin jar of distilled water); dip once to wash away excess blue stain.
15. 95 percent ethanol, twice, 1 minute each.
16. 100 percent ethanol, twice, 1 minute each.
17. Clear in xylene and cover (Permount).

RESULTS

Blue: neuron cell; cytoplasm of interstitial, ectoderm, and gland cells; isorhiza; capsule of desmoneme; granules of endoderm cells.

Pink: granules, cytoplasm, and nuclei of endoderm cells; nuclei of gland and interstitial cells; capsule of stenotele.

SOLUTIONS

Haller's Fluid

Glacial acetic acid	33 cc
Glycerine	33 cc
Distilled water	33 cc

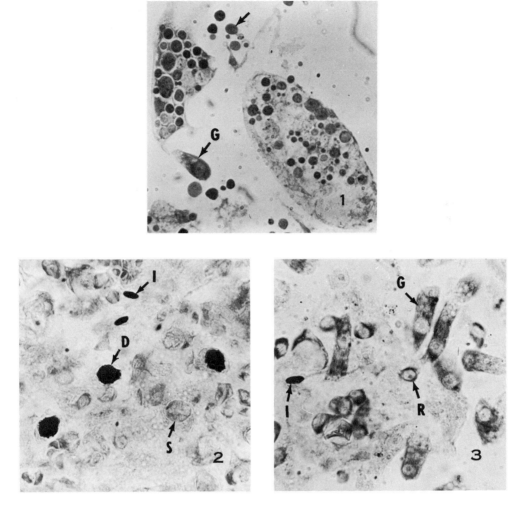

Figure 5–2. Macerated cells of *Hydra.* Methylene blue-phloxine. 750X.

1. Two endoderm cells filled with granules of various colors. Such large cells easily rupture during maceration and spill out the granules (arrow). Gland cell (*G*).

2. Nematocyst types. Isorhiza (*I*) and desmoneme (*D*), deep blue; stenotele (*S*), pink.

3. Gland cell (*G*), blue; reserve cell (*R*), blue; isorhiza (*I*), deep blue.

Phloxine Solution

Phloxine B, C.I. 45410	0.5 gm
Distilled water	100 cc
Glacial acetic acid	0.2 cc

A slight precipitate will form. Filter the solution before you use it.

Methylene Blue Solution

Methylene blue, C.I. 52015	0.25 gm
Azure B, C.I. 43830	0.25 gm
Distilled water	100 cc
Borax	0.25 gm

D. MACERATION OF ONION ROOT TIPS

(Modified by Stamler from Bowen, 1963)

Changes in the chromosome structure and orientation during mitosis are best stud-ied in the root tip squash preparation. It is a simple, rapid method and provides a more accurate picture of mitosis than the more conventional sectioned prepara-tions do.

THE ONION AS A SOURCE OF ROOT TIPS

The best sources of root tips are the small white onions and the onion sets available in almost all grocery stores. Set an onion on a jar of water overnight in the dark. The diameter of the mouth of the jar should be slightly smaller than the diameter of the onion so that the bulb is supported on the rim of the jar and its root base is in contact with the water. Once the roots have started to grow, the preparation can be brought into the light for observation. To harvest the root tips cut the whole root away from the bulb but fix only the distal 5-mm piece. Harvest the tips between 11:30 AM and 1 PM to obtain the maximum number of mitotic figures.

If the small onions are unavailable, use onion seed. Scatter the seeds on moist filter paper in a Petri dish. Keep the Petri dish in the dark for three or four days. Root tips 5 mm long can easily be cut off with a razor blade.

SQUASH PREPARATION

1. Fix in Farmer's solution, 1 hour to overnight. Root tips may be stored in this solution indefinitely.

2. Hydrolyze, 30 minutes. If speed of preparation is desired, freshly cut root tips may be placed directly in hydrolyzing solution without prior fixation.

3. Wash in 95 percent ethanol, 30 minutes minimum with at least 4 changes. It is very important to wash out all the hydrochloric acid.

4. Transfer the root tips to a small Stender dish or to a slide and cover them with Wittman's hematoxylin, 20 minutes.

5. Place a drop of Hoyer's mounting medium on a clean slide.

6. Rinse the tips *briefly* in glacial acetic acid to wash off any excess hematoxylin, blot, and place them in the center of the Hoyer's medium. Let them sit for 30 seconds and then add a coverslip.

7. Hold one corner of the coverslip to prevent it from sliding and gently tap the coverslip repeatedly over the root tip with the blunt end of a pair of forceps or a pencil. When the tip is broken and partially spread, press it with the eraser of a pencil. Finally, cover the slide with a paper towel and press it firmly over the coverslip with your thumb.

8. The slide can be examined under the microscope immediately. Hoyer's medium is permanent as well as water soluble; do not clean the slide with water until the medium has dried for some time.

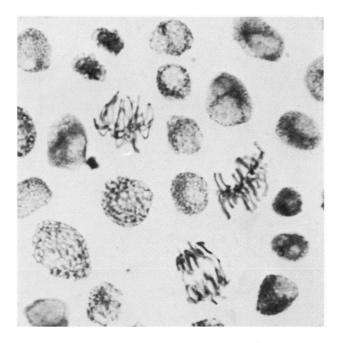

Figure 5–3. Nuclei and mitotic figures of onion root tip squash.

RESULTS

Purple to black: nuclei and chromosomes.

SOLUTIONS

Farmer's Solution

Glacial acetic acid	25 cc
100 percent ethanol	75 cc

Wittman's Hematoxylin

Dissolve: Hematoxylin, C.I. 75290	2	gm
In: Glacial acetic acid	22	cc
Distilled water	28	cc
Add: Ferric ammonium sulfate	0.5	gm
(Iron alum: $Fe(NH_4)(SO_4)_2 \cdot 12 H_2O$)		

Let the solution stand for 24 hours before you use it. The stain solution will keep 4 to 6 months if refrigerated.

Hydrolyzing Solution

Concentrated hydrochloric acid	10 cc
95 percent ethanol	30 cc

Hoyer's Mounting Medium is commercially available from Ward's Natural Science Establishment (see Appendix B).

E. WHOLE MOUNTS OF FROG PARASITES

THE FROG AS A SOURCE OF FLATWORMS

Rana pipiens, the leopard frog, is a good source of parasites. Almost any frog from a commercial supply house will provide several species of flatworms, nematodes, and tapeworms for interesting studies. Free-living flatworms, such as *Planaria sp.,* also are suitable for whole mounts, but they may be so pigmented that their internal organs are obscured. Such specimens must be bleached before they are stained.

1. Kill a frog by pithing.   Using clean scissors and forceps, open the ventral body wall without cutting into the internal organs.   Be careful not to

squeeze and lose the contents of the following organs when you remove and place them in a shallow dish of saline solution: lungs entire, urinary bladder, and rectum.

2. Frog lung fluke (*Haematoloechus*). Carefully open the lung with scissors. It is a simple sac, cut only the wall. Draw up the flukes (flatworms) with a pipette and place them in a separate small dish of saline solution. *Haematoloechus* is a flatworm with an oral sucker. It moves with euglenoidlike movements. If any roundworms (nematodes) are present, they are probably *Rhabdias* (see Hyman, 1951, for further identification).

3. Bladder fluke (*Gorgodera, Gorgodorina*). Carefully open the bladder. Usually there are one or more flukes present. Typically they have a small oral sucker and a larger sucker just anterior to the central portion of the body. Place them in a separate dish of saline solution.

4. Frog rectal fluke (*Megalodiscus*). Carefully open the rectum. *Megalodiscus* is a fluke with a small oral sucker and a large sucker at the posterior end. Place it in a separate dish of saline solution.

PREPARATION OF WHOLE MOUNTS

1. Place a fluke in a small quantity of saline solution on a slide. Withdraw as much fluid as possible, without restricting the fluke's movements. When the worm is extended and straight, place a dropperful of Bouin's fixative directly on it. Allow the fixative to infiltrate the worm for about 5 minutes while you fix a few more worms.

2. Arrange the worms in Bouin's fixative on a slide in the bottom half of a Petri dish. Cover them gently with another slide to flatten them and fill the Petri dish with Bouin's fixative. Allow them to fix 24 to 48 hours. Loosen the top slide occasionally to allow fresh fixative to reach the specimens. The top slide can be removed after the first 24 hours.

3. 50 percent ethanol saturated with lithium carbonate, 1 hour in a covered Stender dish.

4. 70 percent ethanol saturated with lithium carbonate, until all yellow color is gone. If the worms are heavily pigmented (e.g., *Planaria*), soak the specimens in 10 cc 70 percent ethanol with 4 drops of Clorox. Small specimens may bleach in only a few hours. *Planaria* may take overnight.

5. 50 percent ethanol, 1 hour.

6. 35 percent ethanol, 1 hour.

7. Distilled water, 1 hour.

8. Carmalum, overnight.

9. Dehydrate to 70 percent ethanol, 1 hour each.

10. 1 percent hydrochloric acid in 70 percent ethanol; destain until the worm is clear pink and translucent, 2 hours to overnight.

11. 70 percent ethanol with lithium carbonate, 1 hour or until ready to proceed.

12. 95 percent ethanol with lithium carbonate between two slides to ensure flatness, 1 hour.
13. 100 percent ethanol between two slides, 1 to 2 hours.
14. Fast green, 15 to 30 seconds (30 seconds for *Planaria*). Flukes should have mainly pink features.
15. 100 percent ethanol : cresote (beechwood) : xylene (2:1:1), 1 hour. Use a glass container *in the hood*. Creosote is very caustic so keep it off your skin and clothes. It is used to ensure complete dehydration.
16. Creosote : xylene (1:1), 1 to 2 hours to overnight. Pour the used creosote mixtures into a *waste* creosote jar.
17. Fresh creosote : xylene (1:1), 1 hour.
18. Pour off enough creosote mixture to leave the specimens barely covered. Add two drops of Permount every hour or so over a 24-hour period. Try to reach about 90 percent Permount slowly in a minimum of 24 hours. Delicate specimens cannot stand sudden changes in viscosity.
19. Mount directly from the foregoing mixture. If the flukes are very thick, slip several pieces of broken glass slide or two strands of monofilament fishing line under the coverslip to brace it. You may have to add more Permount as the slide dries. Use thick Permount.

SOLUTIONS

Bouin's Fixative

Carmalum

Carminic acid, C.I. 75470	1 gm
Ammonium alum ($NH_4 Al(SO_4)_2 \cdot 12 H_2O$)	10 gm
Distilled water	200 cc
Filter. Add	
Formalin	1 cc

Fast Green

Fast green FCF, C.I. 42053	0.2 gm
100 percent ethanol	100 cc

Saline Solution

Sodium chloride	0.7 gm
Distilled water	100 cc

F. MACERATION OF WOOD FIBERS

MACERATION OF WOOD

1. Use a razor blade to shave off splinters from a small block or board of one of the softer woods (e.g., white pine). The splinters should be smaller than a toothpick.
2. Place the splinters in a glass container (with a glass cover) half filled with Emig's macerating fluid in a 50°C oven until the splinters look fuzzy and whole cells break away (2 hours for white pine).
3. Carefully pour off the acid and wash the fibers with several changes of tap water. Fill the container about half full of tap water and cover it tightly. Shake the container vigorously to separate as many cells as possible. Shake with glass beads in the container *only* if the splinters do not break up easily.
4. Wash the fibers with several changes of tap water over a period of 24 hours. Some of the cells are very small and will be washed away if solutions are changed by simple decanting. Use of a centrifuge is advisable. A simple specimen carrier can be made out of 1/2-inch of glass tubing with one end covered with fine mesh cloth. Be careful to choose a cloth that is insoluble in creosote-xylene (e.g., nylon or silk). A Gouch crucible with a small piece of filter paper in the bottom or a small funnel with filter paper that will rest upright in a small beaker and can be covered by a Petri dish also makes a good specimen holder.

STAINING

5. Dehydrate the fibers to 50 percent ethanol, 5 minutes each.
6. Safranin (dilute stock with distilled water, 1:1), 1 hour.
7. Rinse briefly with 70 percent ethanol.
8. 100 percent ethanol, two times, 1 minute each.
9. Xylene, three times, 5 minutes each.
10. Place a drop of Permount in the center of a clean slide. Pick up a small bunch of fibers with forceps and place them on the Permount. If they do not spread satisfactorily, tease them apart. Cover the slide.

SOLUTIONS

Emig's Macerating Fluid (Emig, 1959)

Chromium trioxide (CrO_3)	10 gm
Distilled water	90 cc
Nitric acid (HNO_3)	10 cc

Safranin

Safranin O, C.I. 50240	1 gm
95 percent ethanol	100 cc

For use, dilute with distilled water (1:1).

Chapter 6

THE MICROSCOPE

A. OPTICS

Light is made up of electromagnetic waves that have definite dimension, that travel in a straight line, and that can be bent in an angle. The speed of light varies according to the medium through which the light beam is traveling. The refractive index of the medium is the ratio of the speed of light in a vacuum to the speed of light in that particular medium. The refractive index of air is 1.00029, whereas the refractive index of glass (lenses, slides, coverslips) is approximately 1.5. If a beam of light passes from a medium of low refractive index (e.g., air, 1.00029) to a medium of higher refractive index (e.g., glass, 1.5), the velocity will change causing the direction to change and thus the light beam is bent or "refracted" (Fig. 6–1).

Microscopes usually have convex lenses that utilize the properties of light refraction and cause a beam of light to converge or diverge according to the design of the lens. The image of any object that stands in the light path as the light travels to and through a convex lens also depends on the design of the lens and the distance of the object from the lens. If the distance from an object to a convex lens is *greater* than the distance of the point of focus (F) to the lens, then the image formed by the lens will be real, inverted (upside down), magnified, and formed on the opposite side of the lens. An image is "real" if it can be seen formed on a screen or sheet of paper that is placed in the path of light at the point at which the rays that passed through any one point on the object converge to form the image. However, if the distance from an object to a convex lens is *less* than the distance of the point of focus (F) to the lens, then the image is virtual, erect (right side up), magnified, and formed on the same side of the lens as the object. An image that is "virtual" cannot be intercepted by any screen. It does not really exist in space. It is an illusion formed by the brain after the diverging rays of light have been gathered by the cornea and lens of the eye. The value of the convex lens in the microscope lies in its ability to place the position of the image quite precisely on either side of the lens and to magnify it under both conditions.

REFRACTION OF LIGHT

i ANGLE OF
   INCIDENCE
r ANGLE OF
   REFRACTION

REFRACTIVE INDEX

WATER   = 1.33
AIR   = 1.00029
GLASS   = 1.5
PERMOUNT   = 1.5
IMMERSION OIL = 1.5

PRISM DISPERSION

CONVEX LENS

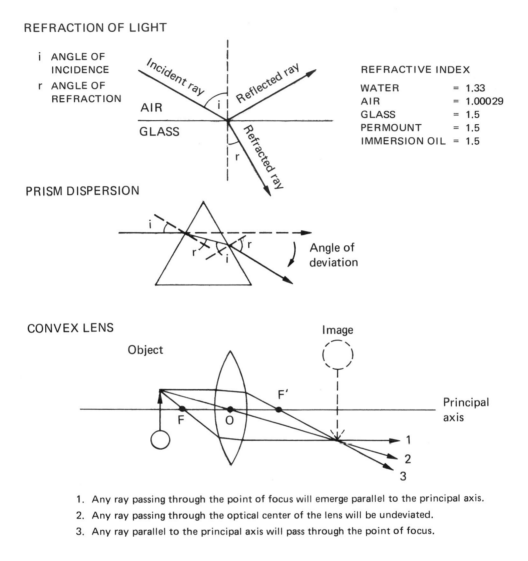

1. Any ray passing through the point of focus will emerge parallel to the principal axis.
2. Any ray passing through the optical center of the lens will be undeviated.
3. Any ray parallel to the principal axis will pass through the point of focus.

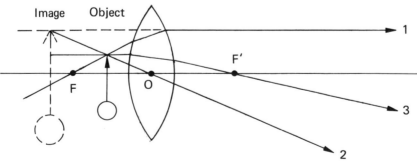

Figure 6-1. Diagrams of geometrical optics.

B. OPTICS OF THE LIGHT MICROSCOPE

The bright-field light microscope is made up of four basic lens systems. The condenser focuses the light precisely on the object on the slide. The objective produces a magnified image of the object on the slide. Quality control of the final microscope image is primarily controlled by the quality of the objective. The ocular magnifies the objective image and presents a final image that the eye translates to the brain (Fig. 6–2).

Information concerning optical characteristics is usually engraved on the side of each objective. For example, on a typical objective

Plan 40/0.65
160/0.17

These figures refer to

Correction type   Initial magnification/Numerical aperture
Tube length/Coverslip thickness

This objective is a planachromat. It magnifies the object 40 times (linear magnification); its numerical aperture is 0.65; it is designed for a microscope having a tube length of 160 mm and is adjusted to be used with coverslips that are 0.17 mm thick ($\#1\frac{1}{2}$).

Aberrations, defects in the image-forming powers of a lens, are caused by a lens design that prevents the formation of a precise image. Objectives of varying degrees of correction for aberrations are available. The achromatic objective, which is the standard objective for most routine work, has a minimum correction for aberrations and a relatively long working distance between the front lens and the object on the slide. Increased correction for aberrations in the objective usually causes a concomitant decrease in working distance as additional lens elements are added to the lens system. Working distance becomes an important factor as the front lens of the longer, better quality objective approaches the top of the coverslip. Planachromatic objectives are corrected for curvature of field and are used for photomicrography. Apochromatic lenses are corrected for spherical and chromatic aberrations and produce an image of high resolution. Fluorite objectives contain fluorite elements that provide good contrast with a resolution quality between the achromat and the apochromat.

Magnification is one of the major purposes of a microscope. Total magnification is a function of the magnifying abilities of the objective and the ocular at a particular tubelength. Modern microscopes are constructed with a set tubelength of either 160 or 170 mm. It is important to avoid mixing up objectives designed for one

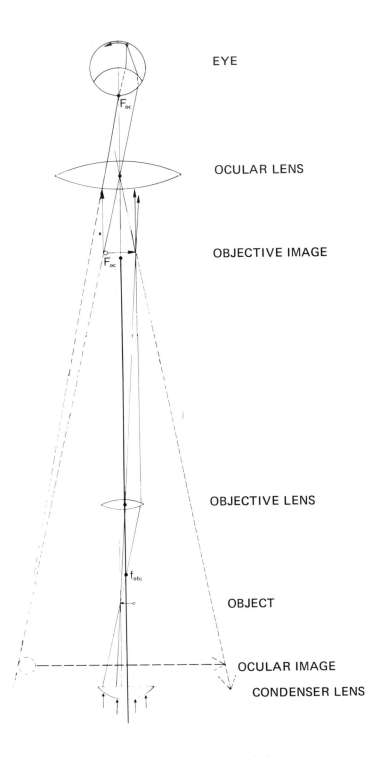

EYE

OCULAR LENS

OBJECTIVE IMAGE

OBJECTIVE LENS

OBJECT

OCULAR IMAGE

CONDENSER LENS

Figure 6-2. Image formation in the light microscope using a single lens system.

tubelength with a microscope designed with a different tubelength. If the tubelength characteristics of objective and microscope match, magnification is calculated by multiplying the initial magnification of the objective by that of the ocular.

A microscope is capable of unlimited magnification. However, magnification is useless unless the magnified image is formed with fine detail or good resolution. If two point objects are so close together that the image formed by the microscope is a single blur, then the magnification is useless because the two objects cannot be resolved as two objects. The minimum separation between two point objects on a slide at which the microscope will still produce an image of two distinct point images determines the resolving power of the microscope. This minimum resolvable distance (h) is a function of the wavelength of light (λ) illuminating the slide and the numerical aperture (NA).

$$h = \frac{0.61\lambda}{NA}$$

The resolution of a microscope is increased by reducing the minimum resolvable distance (h). The light microscope is restricted to the wavelengths of visible light (4000–7000Å). However, the resolvable separation between two point objects can be improved by increasing the numerical aperture.

The numerical aperture (NA) of an objective is an arbitrary number indicating the light-gathering abilities of the objective. It is dependent on the cone of light delivered by the condenser and the refractive index of the medium between the light source (condenser) and the front lens of the objective. Low-powered ("dry") objectives have air between them and the slide with a refractive index of 1.0. All high-powered objectives are used with immersion oil which has a higher refractive index (1.5), and therefore a higher numerical aperture. In general, the total useful magnification of each achromatic objective can be considered to be the product of a thousand times the numerical aperture of that objective.

C. ADJUSTING THE LIGHT MICROSCOPE

BASIC EQUIPMENT FOR A MICROSCOPE

The quality of light microscopy is a direct function of the quality of the microscope and the microscopist. With a slight improvement of equipment and understanding, a microscopist can improve his work manyfold. Unfortunately, many laboratories do not have even minimum basic equipment.

Basic microscope should include:

1. A sturdy microscope stand
2. Oculars (either monocular or binocular)
3. Objectives (low and high power, including an oil immersion high power)
4. A rack and pinion condenser with an aperture diaphragm (some method of controlling substage aperture should be present even if condenser lens elements are absent).
5. An adequate light source of variable intensity (preferably with a field diaphragm and focusing lens).

It should be stressed here that, next to the lenses, the light source is the most important feature of a microscope. Resolution is a function of wavelength and the amount of light entering the front lens of the objective. Because we are currently restricted to visible light wavelengths, the only hope of achieving adequate microscopic observation is to control the light source. This requires an adjustable substage condenser and a light source of variable intensity.

BASIC ADJUSTMENT OF THE LIGHT MICROSCOPE

1. Clean all glass surfaces carefully with lens paper.
2. Open all diaphragms—aperture (condenser) and field (lamp) diaphragms. Place the condenser as high as possible without having it touch the bottom of the slide (or the largest aperture if no condenser available).
3. Always begin by placing the lowest power objective (4X or 10X) in position over the stage.
4. Center the specimen on the slide over the opening in the stage. Hold it in position with clips or a mechanical stage clamp.
5. Watch from the side while you move the objective and the slide together with the coarse adjustment until they almost touch.
6. Raise the microscope tube (or lower the stage) carefully with the coarse adjustment knob until the specimen is in sharp focus when you observe it through the oculars.
7. Look through the ocular. If your microscope is equipped with an external light source and a mirror, adjust the mirror to provide the brightest illuminated field. Without the substage condenser, use the concave side of the mirror; with the condenser, use the flat surface.

ADJUSTMENT OF THE CONDENSER AND THE APERTURES

8. Microscope with no substage condenser and an external light source:

    a. Adjust the concave surface of the mirror to give the best illumination. The lamp should be 10 inches away from the mirror or directly under the substage if no mirror is present.

      b. When you change to objectives of higher magnification, re-adjust the mirror as necessary.

   9. Microscope with substage condenser and a nonadjustable light source:

      a. If the microscope is equipped with a mirror, use the flat surface.
      b. Remove the ocular and observe the back lens of the objective. Open the condenser diaphragm (aperture diaphragm) until the circle of light fills about seven eighths of the tube (field).
      c. Replace the ocular. Adjust the condenser up and down until the field of view is uniformly filled with light.

 10. Microscope with substage condenser and an adjustable light source with field diaphragm and condenser lens (whether built in or external):

      a. Center and focus the light source. If an external lamp is used, it should be about 10 inches away from the mirror. Place a sheet of lens paper or a ground-glass filter over the mirror or the exit pupil of the built-in light source. Center the image of the light source with centering screws or by moving the lamp or mirror. Focus the image of the lamp filament with the lamp condenser or by moving the lamp bulb.
      b. Check through the ocular to make sure that the specimen on the slide is in focus. Close the field (lamp) diaphragm and, while observing through the ocular, adjust the condenser until the edges of the diaphragm are in sharp focus. Open the field diaphragm until it just disappears out of the field of view.
      c. Remove an ocular and close the aperture (condenser) diaphragm until it appears within the field of view or until it closes off all but seven eighths of the field of view. Your microscope is now adjusted for the precise Köhler illumination necessary for maximum resolution.

ADJUSTMENT OF THE LIGHT SOURCE

All changes in light intensity must be made at the lamp. Do not adjust the intensity by changing the setting of the condenser or the condenser diaphragm. If you do not have a variable or step transformer attached to your light source, you can easily reduce the light intensity by placing filters (ground glass, neutral density filters, or sheets of lens paper) in the light path.

CHANGING THE MAGNIFICATION

Always start with the lowest power objective. Never try to focus directly on a fresh slide with any objective higher than 10X. Always go back to low power.

11. When the specimen is in sharp focus under the low-power objective, rotate the nosepiece until the next higher powered objective is in position. This should be done with care the first time you work with any microscope. Not all microscopes have matched objectives to clear the coverslip when the nosepiece is rotated.

12. Use the fine focus adjustment knob to focus carefully on the specimen. Never use the coarse adjustment knob with objectives higher than 10X.

13. Close the field (lamp) diaphragm and focus on the edges by moving the condenser up and down. Open the diaphragm until it is just outside the field.

14. Remove an ocular and adjust the aperture (condenser) diaphragm until it is visible inside the field of view. Replace the ocular.

15. Adjust the light intensity with the lamp transformer or by removing filters or sheets of lens paper from the light path.

OIL IMMERSION OBJECTIVES

The highest magnification objectives must have immersion oil between the front lens of the objective and the coverslip in order to increase the light-gathering ability of the optical system required for high powers. Following the normal practice of oiling only the objective, a condenser with a numerical aperture (NA) of 1.0 is all that is needed. With condensers and objectives of higher NA (1.2 or 1.3), oil should be placed between the top lens of the condenser and the bottom of the slide as well as under the objective for maximum resolution.

16. Start with the low-power objective. Increase the magnification stepwise until you reach the highest powered dry objective (generally 40X). Center the specimen carefully and focus with the fine adjustment knob.

17. Rotate the nosepiece to a position about halfway between the high-dry objective and the oil immersion objective. Place a drop of oil on the coverslip centered in the ring of light coming up from the condenser. The oil drop must be free of bubbles.

18. Carefully rotate the nosepiece to bring the front lens of the immersion objective into contact with the oil. Watch the objective from the side to prevent the front lens from hitting the coverslip and being scratched.

19. Use *only* the fine adjustment knob with oil immersion objectives. Focus carefully on the specimen, adjust the aperture and field diaphragm as before (13, 14), and adjust the light intensity (15).

20. To oil the condenser, lower it until you can place a drop of oil on the top lens. The oil must be free of bubbles. Raise the condenser until the oil comes in contact with the bottom surface of the slide. Again check for bubbles.

21. Readjust the diaphragms as before.

Oil immersion objectives must always be cleaned carefully after they are used. Wipe off any excess oil with lens paper. Wipe the surfaces of the lenses lightly with lens paper moistened with xylene. Immediately wipe off the xylene with clean lens paper. All oil should be wiped off the stage before you put the microscope away.

D. FUNCTION OF THE OBJECTIVE AND OCULAR

OBJECT

Choose a number near the center of a calendar slide (a letter "e" slide is also satisfactory). The number should be asymmetrical in shape so that you can easily determine whether the image is right side up (erect), upside down (inverted), or has been turned over laterally. Measure the length of the number on the slide with a millimeter rule. This is the size of the object to be magnified by the microscope.

OCULAR

Remove the ocular from the microscope tube. Unscrew the top lens (eye-lens) and cover the tube to keep dust from the prism. Place the slide on a brightly lighted sheet of white paper. Hold the eye-lens about 1 inch above the slide and keep your eye about 10 inches above the eye-lens. As you raise the eye-lens above the slide, the image will enlarge until it blurs. Raise the eye-lens to a point just before it blurs. In this position, the number object is located between the lens and the equivalent point of focus. Note that the image is magnified and erect. Place a millimeter ruler on the paper near the slide and estimate the size of the image in millimeters.

Place a ground-glass filter directly on the top of the ocular. Focus your eyes on the filter; raise and lower the filter above the ocular and note that at no distance above the ocular is a sharp image formed on the ground glass. The image is virtual.

Raise the eye-lens through the distance at which the image is blurred until the image is again sharp (about 2 inches). Note how the image has changed. The object is now located outside the equivalent point of focus. It is magnified and inverted. Again place the ground-glass filter directly on top of the ocular. Note that a sharp image (magnified and inverted) is formed on the ground glass approximately 3 cm above the ocular. Therefore, the image is real. Note the conditions under which the ocular can produce two kinds of image.

OBJECTIVE

Place the slide on the stage of your microscope under the lowest power objective and with the reassembled ocular in place. Focus carefully and note the orientation of your selected number object. Remove the ocular and increase the illumination. Place the ground-glass filter on the upper end of the microscope tube and focus the objective image on the ground glass. Measure the length of the objective image with a millimeter rule. Note that the image is inverted, magnified, and (because it can be formed on the ground-glass surface) real. Replace the ocular. Note that both images, with and without ocular, are inverted.

THE MICROSCOPE (OBJECTIVE AND OCULAR TOGETHER)

Focus carefully through the re-assembled microscope and note that the image is inverted, magnified, and virtual (consult Fig. 6–2). The virtual image is set to look as if it were located 10 inches away because this is the standard near-point distance for the eye.

If the objective image is inverted and real, what is the function of the ocular? Remember that the objective image is the ocular object. In a microscope, where is the objective image formed in respect to the ocular lenses? On what evidence do you base your answer?

Hold a millimeter rule in front of the stage approximately 10 inches from the eyepoint in a line with the sloping ocular tube. Look with your weaker eye through an ocular and with your stronger eye at the millimeter markings on the rule. Try to superimpose the two images. You may have to move the object on the slide slightly in order to bring the images together. Estimate the length of the ocular image by reading the ruler.

Compare the magnification by the objective with that of both the objective and ocular together. Be sure you understand in what ways the objective and ocular contribute to the total microscope image as seen by the eye at the eyepoint.

E. INSTRUCTIONS FOR A DRAWING

The purpose of a drawing is to provide an accurate picture of an object based on careful observation. It must convey a clear idea of the actual structure as well as the correct relationship of its parts. The essential characteristics of a good drawing are clarity, complete labeling, neatness, and accuracy.

The drawing should be centrally located on a page of good quality white notebook paper. The complete title, which includes magnification, should appear beneath the drawing. Place your name and date in the upper right-hand corner of the page.

Work neatly and cleanly. Use a well-sharpened pencil, preferably 4H or 5H, and keep it sharp. Draw each line with a single stroke. Do not make sketchy lines. Shading should not be used.

Make the drawing big enough to be seen and to be labeled easily. In general, place only one or two drawings on a page.

Guidelines should connect all structures with their labels. The lines must be straight and continuous. They must not cross each other. The labels, to facilitate reading, should be printed and parallel with the top of the page. If possible, place all labels on the same side of the drawing, preferably the right side, and try to keep them all approximately the same distance from the picture.

A scientific drawing is not an expressionistic piece of art. It is, we hope, a simple impression that will convey the essential details to another observer. Select as typical a specimen as possible. The importance of the drawing is more in observation than in technique.

MAGNIFICATION

Always indicate the magnification of the drawing by printing directly beneath the title a scale line stating length. Determine the scale of your drawing by measuring the specimen directly with an ocular micrometer or with the mechanical stage vernier.

The ocular micrometer, a disc with a scale engraved on it, can be inserted into the ocular. Naturally, each unit of the micrometer will represent a different scale with each different objective or change in magnification. The ocular micrometer is calibrated for each objective by focusing on a scale of known dimensions on the stage of the microscope. Calculate the distance represented by one unit of the micrometer for each objective. If a stage micrometer is not available, use a hemacytometer. On the AO Spencer hemacytometer, the side of one of the smallest squares is 0.05 mm and a group of 16 smallest squares is bounded by triple lines which are 2.5 microns apart.

The vernier scale on the mechanical stage provides a convenient method for measuring a specimen. The specimen may be measured directly or the field of view can be measured and the ratio of the diameter of the specimen to that of the whole field can be estimated. If the specimen is to be measured directly, it should be observed under a magnification by which it fills as much of the field of view as possible. Move the slide back and forth using a single vernier screw knob. After noting the direction of movement of the specimen, place one end of the specimen at the edge of the field of view along the line of travel of the slide and write down the reading of the vernier scale of the knob you are using. Move the specimen until its opposite side rests at the same edge of the field of view. Note again the vernier reading. The difference between the two readings will represent the diameter of the specimen in

millimeters.   The difficulty with this method is that the minimum measurable distance is 0.1 mm.   For smaller specimens it is best to measure the diameter of the
field and then estimate the diameter of the specimen from that figure.

The scale line on your drawing should measure between 1 and 2 inches long. It
should be drawn to represent 0.1, 0.01, or 0.001 mm or a convenient decimal to fit
a line about an inch long.   The scale line is a more accurate method of recording
magnification than the simple multiplication of the magnifications of the ocular and
the objective.

Chapter 7

SLIDE ANALYSIS

A. CELL IDENTIFICATION IN DIFFERENT TISSUES

It is essential that you, as a student of microtechnique, learn to judge slides accurately as well as make them. This can be done only if you study the slides you have made during the course. You must have a well-adjusted microscope and a knowledge of the tissues you have prepared. Go over your slides carefully. Familiarize yourself first with the tissue, then with the stain reaction, and then learn to recognize the many artifacts that make reading a slide such a complex and skilled process. At first, try following these steps.

1. Familiarize yourself with the relative position and shape of each cell type in each tissue. Be sure you know the general function of the organs represented on your slides. Take advantage of the references available in the laboratory.
2. Note specific stain reactions of each cell type. You have used a number of stain combinations. What information does each stain provide? When you have examined all stain reactions of a cell type, what prediction can you make concerning its function?
3. Correlate the structure, stain reactions, and position of each major cell type and try to relate it to the organ function.
4. Be aware of the differences in stain reaction that result from differences in processing. Do the duodenal epithelial cells stain with phloxine-methylene blue in the same way after Bouin's fixation as after Carnoy's fixation? How does the staining difference affect the interpretation of the slide preparation? Which is the best fixative? What are your criteria? Is that space around the nucleus in the hepatic cell really there or does it occur because of shrinkage? What are the shrinkage characteristics of the fixatives you used?
5. Finally, it is important to be able to recognize, and thereby to discount, the ubiquitous air bubble, dirt particle, dandruff flake, stain precipitate, water droplet, and black mercury deposit left over from Helly's fixative. Now is the time to recognize how slide processing can go wrong so that in the future when you are working on very important and perhaps irreplaceable tissues you will have the experience to handle the material correctly and intelligently.

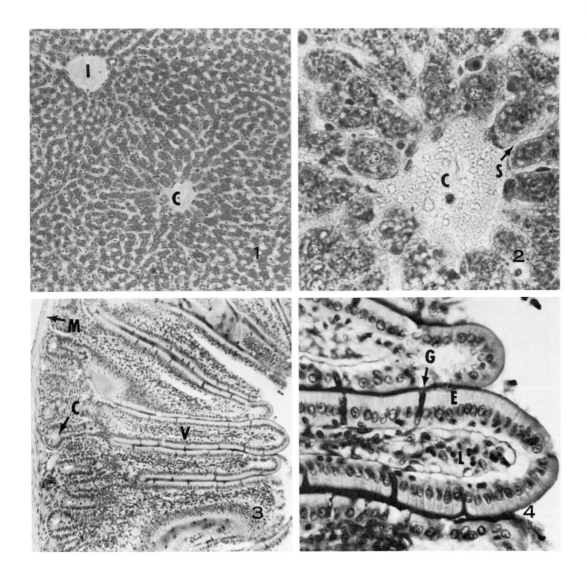

Figure 7-1.

1. Mouse liver. Central vein (C); interlobular vein (I). Carnoy, methylene blue-phloxine. 140X.

2. Mouse liver. Central vein (C); sinus (S). Carnoy, methylene blue-phloxine. 500X.

3. Mouse intestine. Villus (V); muscularis mucosa (M); crypt of Lieberkuhn (C). Carnoy, periodic acid-Schiff and hematoxylin. 140X.

4. Mouse intestine. Goblet cell (G); epithelial cell (E); lamina propria (L). Carnoy, periodic acid-Schiff and hematoxylin. 500X.

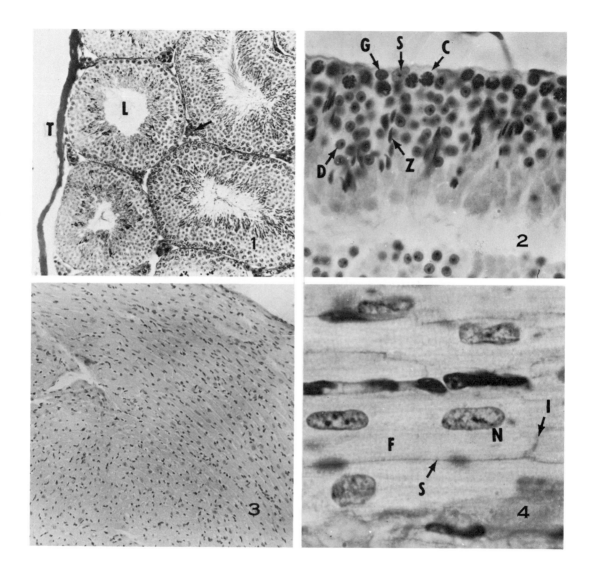

Figure 7–2.

1. Mouse testis. Lumen of seminiferous tubule (L); tunica (T). Carnoy, Feulgen-fast green. 140X.

2. Mouse testis. Spermatogonium (G); Sertoli cell (S); spermatocyte (C); spermatozoan (Z); spermatid (D). Carnoy, Feulgen-fast green. 500X.

3. Mouse heart. Carnoy, periodic acid-Schiff and hematoxylin. 140X.

4. Mouse heart. Muscle fiber (F); sarcolemma (S); nucleus of muscle cell (N); intercalated disc (I). Carnoy, periodic acid-Schiff and hematoxylin. 500X.

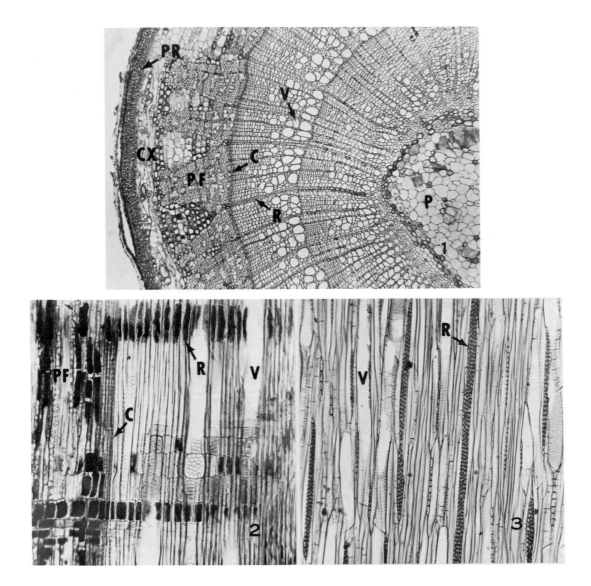

Figure 7–3. *Tilia* twig. FAA, safranin-fast green.

1. Cross section. Cambium (*C*); cortex (*CX*); pith (*P*); phloem fibers (*PF*); periderm (*PR*); vascular ray (*R*). 40X.

2. Radial section. Cambium (*C*); phloem fibers (*PF*); vascular ray (*R*); xylem vessel (*V*). 200 X.

3. Tangential section. Vascular ray (*R*); xylem vessel (*V*). 200X.

B. ULTRASTRUCTURE OF THE CELL

The purpose of biological microtechniques is to prepare a tissue with its component cells and extracellular material so that it is suitable for subsequent microscopic investigation. However, interpretation of the resulting observations requires some knowledge of the ultrastructure of the prepared cells as seen through the electron microscope. It would be convenient to refer here to the "typical" cell. Unfortunately, no such cell exists and each type must be studied for itself. However, certain organelles have been observed in many cells and are considered a common property of all cells of the higher organisms of the plant and animal kingdoms.

The *cell membrane (CM)* is the all-important barrier between the inner and outer environments of the cell. It protects the cytoplasm and controls communication with the extracellular medium. It is often seen through the electron microscope (Fig. 7-7, *2*) as two dense parallel lines enclosing an electron-lucid space.

The *nucleus (N)* is separated from the cytoplasm by a double-membraned *nuclear envelope (NE)* (Fig. 7-5, *1*). The *nucleolus (NL)* contains a high concentration of RNA and is very large in cells that actively synthesize protein (Fig. 7-4, *1*). Flecks of *chromatin (C)* contain high concentrations of DNA.

The endoplasmic reticulum may be smooth or lined with *ribosomes (R)* (Fig. 7-5, *1*). The *agranular endoplasmic reticulum (AER)* is associated with glycogen synthesis in the liver. The *granular endoplasmic reticulum (GER)* is associated with protein synthesis. The *Golgi apparatus (GA)* is associated with secretion (Fig. 7-7, *1*) and formation of the *acrosomal granule (AG)* in the differentiating spermatid (Fig. 7-8, *1*).

The *mitochondria (M)* contain the enzymes of oxidative metabolism (Fig. 7-5, *1*; Fig. 7-6, *1*; Fig. 7-8, *1*). They are the primary sites for the mobilization of utilizable energy *(ATP)* for the cell.

Each cell possesses certain common elements of structure as well as unique features characteristic of each cell type. Heart muscle is made up of highly specialized cells (Fig. 7-5, *2*) which are filled with *myofilaments (MY)* and attached (held together) by cell junctions, *intercalated discs (IC)*. The *sarcomeres (S)* are muscle cell units delimited by the Z lines. Epithelial cells of the duodenum have *microvilli (MV)* for the increase of surface area of the absorbing surface (Fig. 7-7, *2*). The nucleus of the spermatozoan is densely filled with DNA (Fig. 7-8, *2*), and the *acrosomal cap (A)* is associated with mechanisms for the penetration of the egg membrane by the spermatozoan.

The tissues prepared for the electron microscope were fixed in 6.25 percent glutaraldehyde buffered with 0.1 M s-collidine or 0.15 M phosphate buffers at pH 7.4 and postfixed in 1 percent OsO_4 with 4.5 percent sucrose and buffered at pH 7.4 with the same buffers. They were embedded in Epon-Araldite and examined on a Hitachi (HU 11-B-2) electron microscope.

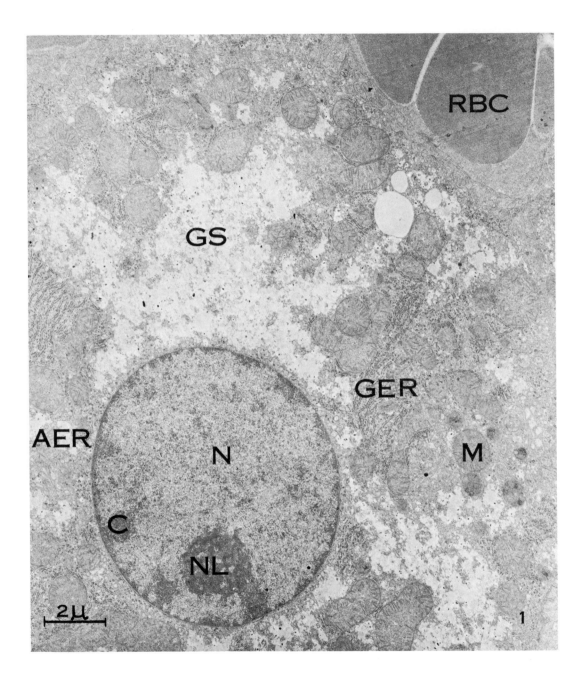

Figure 7–4. Mouse hepatic cell. Agranular endoplasmic reticulum (*AER*); chromatin (*C*); granular endoplasmic reticulum (*GER*); site of glycogen storage (*GS*); mitochondrion (*M*); nucleus (*N*); nucleolus (*NL*); red blood cell (*RBC*). 9750X.

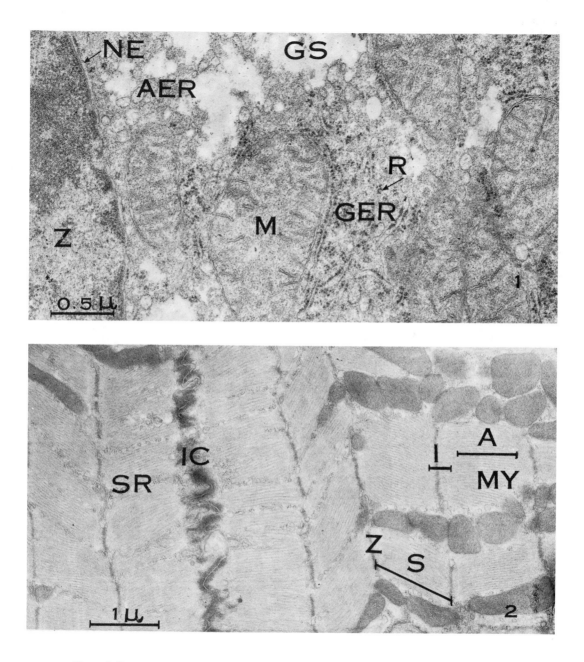

Figure 7-5.

1. Mouse hepatic cell. Agranular endoplasmic reticulum (*AER*); granular endoplasmic reticulum (*GER*); site of glycogen storage (*GS*); mitochondrion (*M*); nuclear envelope (*NE*); ribosome (*R*). 38,800X.

2. Mouse heart muscle fiber (longitudinal section). Anisotropic band (*A*); isotropic band (*I*); intercalated disc (*IC*); myofilament (*MY*); sarcomere (*S*); sarcoplasmic reticulum (*SR*); Z-line (*Z*). 21,000X.

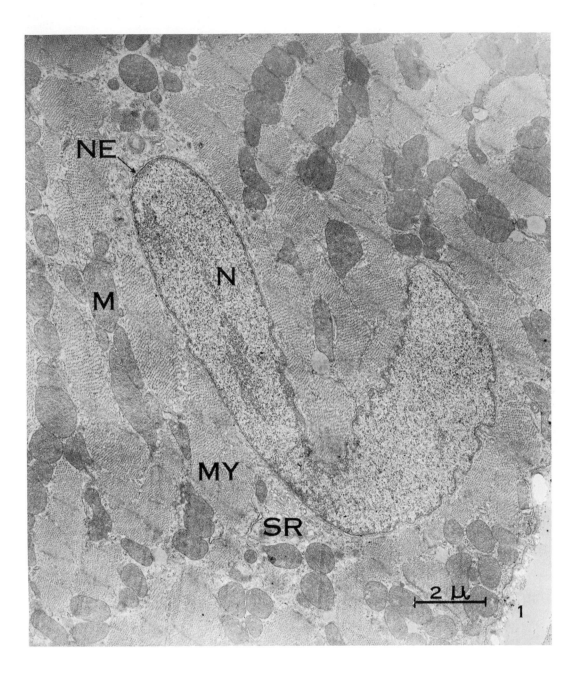

Figure 7-6.　Mouse heart muscle (cross section).　Mitrochondrion (*M*); myofilaments (*MY*);
nucleus (*N*); nuclear envelope (*NE*); sarcoplasmic reticulum (*SR*).　10,900X.

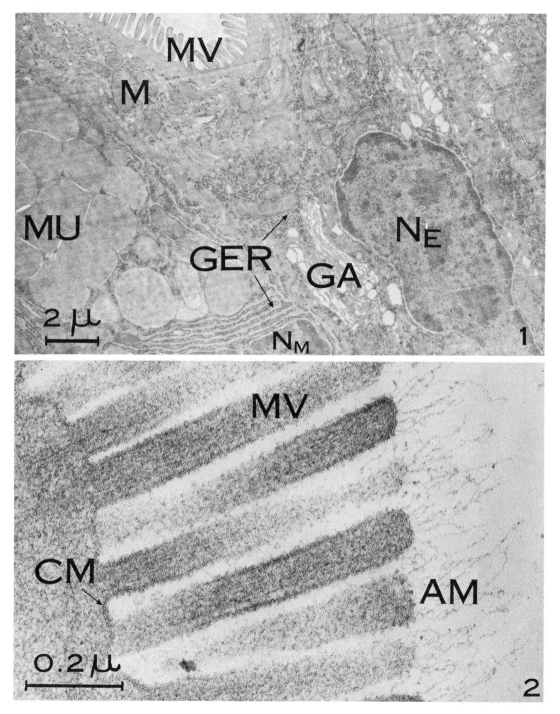

Figure 7–7.
1. Mouse duodenal epithelium. Golgi apparatus (*GA*); granular endoplasmic reticulum (*GER*); mitochondrion (*M*); mucin of goblet cell (*MU*); microvilli (*MV*); nucleus of goblet cell N_M; nucleus of epithelial cell N_E. 8000X.
2. Mouse duodenal epithelium. Antennuli microvillares (*AM*); cell membrane (*CM*); microvillus (*MV*). 150,000X.

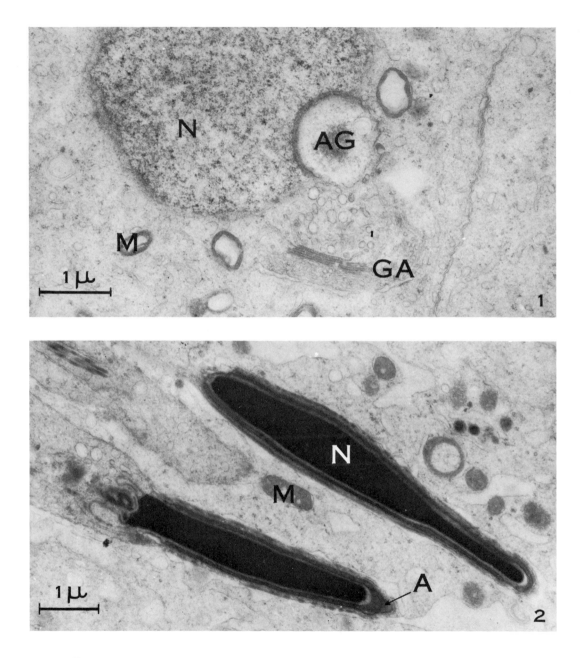

Figure 7–8.

1. Mouse spermatid. Acrosomal granule (*AG*); Golgi apparatus (*GA*); mitochondrion (*M*); nucleus (*N*). 22,200X.

2. Mouse spermatozoan head (longitudinal section). Acrosome (*A*); mitochondrion (*M*); nucleus (*N*). 18,400X.

C. SLIDE ANALYSIS QUESTIONS

1. Compare the effect of fixation in one of your tissues (testis works well) on the staining behavior of methylene blue-phloxine. You should include Bouin's, Helly's, and Carnoy's fixatives.

2. Find a typical hepatic cell. Describe its contents by using your slides and the electron micrographs. State clearly the fixative, stain, and color and use your slides as evidence. What evidence of structure is presented by your slides but not by the electron micrographs? By the electron micrographs, but not by your slides? On the basis of the hepatic cell's gross and chemical structure, what can you say about its function?

3. Spermatogenesis is the differentiation of the spermatogonium into a motile sperm cell. Compare the structures of these two kinds of cells in reference to their functions. Give specific stain reactions as evidence.

4. Many epithelial cells exhibit polarity of structure—the side of the cell attached to the underlying tissue is different from the side located at the free surface. How does this affect the function of the cell? Discuss how this characteristic is exhibited by the epithelium of the duodenum. Use the evidence from your own slides.

5. Polarization is not typical of liver cells. Account for its presence in your slides of liver. It is generally best seen with slides stained with PAS. Why?

6. Compare the effect of the embedding matrix on the quality of a slide preparation for the study of a specific tissue or organ. Under what conditions is each matrix used? Clarify your answer with specific examples from your slides.

Appendix A

GENERAL REMARKS
ON
PRIMARY FIXATIVES

(Modified from Baker, 1958)

ETHANOL

Coagulates cytoplasm as coarse mesh.
Destroys mitochondria.
Tends to fuse and destroy lipid globules.
Shrinks nucleolus.
Leaves chromosomes indistinct.
Causes considerable shrinkage and contraction.
Compatible with picric acid, mercuric chloride, formaldehyde, and acetic acid.
Tends to be oxidized to acetic acid. Avoid mixtures with chromium trioxide, potassium dichromate, and osmium tetroxide.

PICRIC ACID

May be explosive, must be kept damp.
Coagulates nuclear sap.
Preserves chromosomes well.
Leaves cytoplasm homogeneously fixed, badly shrunken, but soft.
Very compatible with other fixatives.

MERCURIC CHLORIDE

Lethal. In small quantities causes acute nephritis.
Preserves cytoplasmic inclusions, e.g., mitochondria and polymorphic neutrophil granules.
Leaves nucleolus very distinct. Chromosomes poorly fixed.
Leaves cytoplasm homogeneously fixed but badly shrunken.
Distorts the cell less than any other fixative.
Produces black artifacts that must be removed by action of iodine in alcoholic solution.
Leaves tissues more receptive to stains than other fixatives do.

CHROMIUM TRIOXIDE

When dissolved in water, becomes chromic acid.
Must be washed in running water because alcohol also aids reduction.

Should be used in the dark; the fixative is light unstable; if it is left for long in light
it reduces to green chromic oxide and is very insoluble.
Excellent for chromosomes and neurons.
Does not fix mitochondria.
Liable to leave cytoplasm coarsely coagulated.
Incompatible with reducing fixatives such as formaldehyde and ethanol.

FORMALDEHYDE

A gas; formalin is 37 to 40 percent formaldehyde in water solution.
Preserves mitochondria well, also protects them from the action of acetic acid.
A poor fixative for paraffin technique; it is better used for frozen or celloidin sections.
Does not protect from extreme shrinkage and distortion by paraffin.

OSMIUM TETROXIDE

Fixes well in vapor form, especially eyes, nose, and mouth.
Penetrates very slowly; only thin material can be used.
Reduces readily in light; must fix in the dark.
Does not protect against shrinkage and distortion caused by paraffin.
Important as a fixative for electron microscopy.
Incompatible with formaldehyde and ethanol.

POTASSIUM DICHROMATE

A poor fixative by itself; it causes too much shrinkage.
Must be washed in running water to prevent reduction to insoluble chromic oxide
by ethanol.
Must be kept above pH 4; below pH 4, ions are the same as chromium trioxide.
Leaves cytoplasm and nuclear sap in homogeneous condition.
Preserves mitochondria well.
Dissolves nucleolus partially.
Leaves chromosomes scarcely visible.
Compatible with picric acid, mercuric chloride, osmium tetroxide.
Reduced by formaldehyde and ethanol to chromic oxide (Cr_2O_3).

ACETIC ACID

Must be kept on acid side of pH 4; above pH 4, tissues macerate with no fixation.
Leaves cytoplasm strongly retracted.
Leaves mitochondria and Golgi apparatus dissolved out.
Fixes nuclear sap poorly.
Fixes chromosomes well.
Compatible with other fixatives; if mixed with potassium dichromate it causes fixa-
tion reaction of chromium trioxide.

COAGULANT FIXATIVES

	Ethanol C_2H_5OH	Picric Acid $C_6H_2(NO_2)_3OH$	Mercuric Chloride $HgCl_2$	Chromic Acid CrO_3
Standard concentration	95–100%	Saturated aqueous 1.2%	Saturated aqueous 6–7%	Aqueous 0.5%
Redox characteristic	Reducer	Oxidizer	Strong oxidizer	Strong oxidizer
Reaction: protein	Strong coagulant; nonadditive	Coagulant; additive	Powerful coagulant	Powerful coagulant; additive
Reaction: nucleoprotein	—	Precipitates protein; leaves DNA in solution	Weak coagulant	Good fixative; coagulant
Reaction: lipids	—	—	Unmask from lipoprotein; no fixation	Good fixative; oxidizer
Reaction: carbohydrates	—	No fixation; binds glycogen to protein	Good for mucopolysaccharides	Oxidizer; changes but no fixation
Rate of penetration	Rather fast	Rather slow	Moderate	Slow
Shrinkage	Strong	Strong; especially after paraffin	Minor	Moderate
Hardening	Extreme	Leaves very soft	Moderate	Moderate
Effect on staining	Little change	Cytoplasm acidophilic	Cytoplasm receptive to basic and acidic stains	Cytoplasm strongly acidophilic
Washing	Alcohol	Water or alcohol	70% ethanol and iodine	Water

NONCOAGULANT FIXATIVES

	Formaldehyde C H_2O	Osmium Tetroxide OsO_4	Potassium Dichromate $K_2Cr_2O_7$	Acetic Acid CH_3COOH
Standard concentration	4% aqueous (10% formalin)	1% aqueous	1.5% aqueous	5% aqueous
Redox characteristic	Reducer	Oxidizer	Oxidizer	Oxidizer
Reaction: protein	Noncoagulant; stabilizes and insoluble in water	Noncoagulant; additive	Noncoagulant; (above pH 4)	Hydrates protein; no fixation
Reaction: nucleoprotein	—	—	Dissolves DNA	Precipitates nucleoprotein
Reaction: lipids	Good preservative	Additive fixative	Good fixative; additive	No fixation; dissolves some lipids
Reaction: carbohydrates	No fixation; traps glycogen	—	—	—
Rate of penetration	Moderate; slow action	Progressively slow	Rapid	Moderately rapid
Shrinkage	Swells slightly during fixation; shrinks after paraffin	Rather little change	Strong shrinkage after paraffin	Swells; strong shrinkage after paraffin
Hardening	Strong	Strong	Very soft	Very soft; prevents subsequent hardening in alcohol
Effects on staining	Cytoplasm basophilic	Cytoplasm basophilic	Variable; good acidophilia	Cytoplasm acidophilic; Chromosomes basophilic
Washing	Water	Water	Water	Alcohol

Appendix B

MISCELLANEOUS FORMULAE

HAUPT'S ADHESIVE

Gelatin (Knox works well)	1 gm
Distilled water	100 cc
Stir and warm gently (not above 30°C). Add	
Glycerin	15 cc
Phenol crystals	1 gm
Stir thoroughly. Filter.	

MAYER'S ADHESIVE

Whip an egg white vigorously. Allow it to stand overnight in the refrigerator. Pour the fluid egg white out from under the stiffened foam. Add an equal amount of glycerine and a crystal of thymol. Mix thoroughly.

HOYER'S MOUNTING MEDIUM

Gum arabic	50 gm
Chloral hydrate	20 gm
Distilled water	50 cc

1 N HYDROCHLORIC ACID

Concentrated HCl (37 to 38 percent assay)	8 cc
Distilled water	100 cc

REFERENCES

Ames Lab-Tek, 1965, *Operating Manual, Tissue-Tek Microtome-Cryostat.* Ames Company, Inc., Westmont, Ill. 34 pp.

Baker, J.R., 1958, *Principles of Biological Microtechnique.* John Wiley and Sons, Inc., New York. 357 pp.

Baker, J.R., 1960, *Cytological Technique.* John Wiley and Sons, Inc., New York. 150 pp.

Bowen, W., 1963, "The demonstration of mitotic figures in green algae," *Iowa Academy of Science,* **70**: 138–142.

Casartelli, J.D., 1965, *Microscopy for Students.* McGraw-Hill Book Co., New York. 138 pp.

Conn, H.J., M.A. Darrow, and V.M. Emmel, 1960, *Staining Procedures.* Williams and Wilkins Co., Baltimore. 289 pp.

Culling, C.F.A., 1963, *Handbook of Histopathological Techniques.* Butterworth's, London. 553 pp.

Drury, R.A.B., E.A. Wallington, and R. Cameron, 1967, *Carleton's Histological Technique.* Oxford University Press, New York. 432 pp.

Dyer, R., and R.L. Willey, 1969, "A technique for preparing permanently stained *Hydra* macerations," *American Biology Teacher,* **31**: 304–306.

Emig, W.H., 1959, *Microtechnique.* W.H. Emig, Colorado Springs. 92 pp.

Emmel, V.M., and E.V. Cowdry, 1964, *Laboratory Technique in Biology and Medicine.* Williams and Wilkins Co., Baltimore. 453 pp.

Galigher, A.E., and E.N. Kozloff, 1964, *Essentials of Practical Microtechnique.* Lea and Febiger, Philadelphia. 484 pp.

Gray, P., 1954, *The Microtomist's Formulary and Guide.* Blakiston Co., New York. 794 pp.

Gray, P., 1958, *Handbook of Basic Microtechnique.* McGraw-Hill Book Co., New York. 302 pp.

Gurr, E., 1960, *Encyclopedia of Microscopic Stains.* Williams and Wilkins Co., Baltimore. 498 pp.

Guyer, M.F., 1953, *Animal Micrology.* University of Chicago Press, Chicago. 327 pp.

Holmes, W., 1947, "The peripheral nerve biopsy." *In* Dyke, S.C., *Recent Advances in Clinical Pathology.* Blakiston, Philadelphia. pp. 404–405.

Humason, G.L., 1967, *Animal Tissue Techniques.* W.H. Freeman and Co., San Francisco. 569 pp.

Hyman, L.H., 1940, *The Invertebrates,* "Protozoa Through Ctenophora," Vol. 1, McGraw-Hill Book Co., New York. 726 pp.

Hyman, L.H., 1951, *The Invertebrates,* "Platyhelminthes and Rhynchocoela; the Acoelomate Bilateria," Vol. 2, McGraw-Hill Book Co., New York. 550 pp.

Jensen, W.A., 1962, *Botanical Histochemistry.* W.H. Freeman and Co., San Francisco. 408 pp.

Johnson, B.K., 1960, *Optics and Optical Instruments,* Dover Publications, Inc., New York. 224 pp. (Reprint of *Practical Optics,* 1947, Constable and Co., London.)

Leitz GMBH, Ernst, 1964, *The Microscope and Its Application.* Ernst Leitz GMBH, Wetzlar. 38 pp.

Martin, L.C., and B.K. Johnson, 1966, *Practical Microscopy.* Blackie and Son, Ltd., London. 138 pp.

Möllring, F.K., 1967, *Microscopy from the Very Beginning.* Carl Zeiss, Oberkochen. 64 pp.

Noller, C.R., 1958, *Chemistry of Organic Compounds.* W.B. Saunders Co., Philadelphia. 978 pp.

Pantin, C.F.A., 1948, *Notes on Microscopial Technique for Zoologists.* Cambridge University Press, Cambridge. 77 pp.

Richards, O.W., 1958, *The Effective Use and Proper Care of the Microscope.* American Optical Co., Buffalo. 63 pp.

Richards, O.W., 1959, *The Effective Use and Proper Care of the Microtome.* American Optical Co., Buffalo. 92 pp.

Sass, J.E., 1961, *Botanical Microtechnique.* Iowa State University Press, Ames. 228 pp.

Sharma, A.K., and A. Sharma, 1965, *Chromosome Techniques.* Butterworth's, Washington, D.C. 474 pp.

INDEX